GRAPHING CALCULATOR
CALCULUS WORKBOOK
AN EXPLORATORY APPROACH

D1652060

YO-DWO.907

GRAPHING CALCULATOR CALCULUS WORKBOOK
AN EXPLORATORY APPROACH

AL SHENK
University of California, San Diego

 HarperCollins*CollegePublishers*

Sponsoring Editor: George Duda
Project Editor: Ann-Marie Buesing
Design Administrator: Jess Schaal
Cover Design: Terri Design/Terri Ellerbach
Production Administrator: Randee Wire
Printer and Binder: Malloy Lithographing, Inc.

TI-81, TI-82, and TI-85 are registered trademarks of Texas Instruments Incorporated; fx-7000G and fx-7700G are registered trademarks of Casio Incorporated; EL-9200 and EL-9300 are registered trademarks of Sharp Electronics Corporation.

Typeset by AMS-TEX

Graphing Calculator Calculus Workbook: An Exploratory Approach

Copyright © 1994 by Norman Al Shenk

All rights reserved. Printed in the United States of America. No part of this book may be used or reproduced in any manner whatsoever without written permission, except in the case of brief quotations embodied in critical articles and reviews. For information address HarperCollins College Publishers, 10 East 53rd Street, New York, NY 10022.

Library of Congress Cataloging-in-Publication Data

Shenk, Al
    Graphing calculator calculus workbook: an exploaratory approach/
Al Shenk.
        p.   cm.
    Includes index.
    ISBN 0-06-501724-2
    1. Calculus – Graphic methods – Data processing. I. Title.
    QA303.5.D37S47 1993
        515' .0285' 41--dc20   93-20906
                                                                    CIP

93 94 95 96 9 8 7 6 5 4 3 2 1

GRAPHING CALCULATOR CALCULUS WORKBOOK
AN EXPLORATORY APPROACH

Preface .... p. v

To the student .... p. ix

Course outline I, for engineering, science, and mathematics majors .... p. xi

Course outline II, for biology, economics, humanities, social science majors ..... p. xvii

PART 1. GRAPHS, FUNCTIONS, AND LIMITS

MAXIMUM/MINIMUM PROBLEMS: INITIAL INVESTIGATIONS

Worksheet 1A.1. Maximizing an area .... p. 1

Worksheet 1A.2. Minimizing a length .... p. 3

Worksheet 1A.3. Minimizing and maximizing an area .... p. 5

Worksheet 1A.4. Finding the closest point on a curve .... p. 7

GRAPHING CALCULATOR INSTRUCTIONS*

Calculator instructions 1B.1. Basic operations .... p. 9

Calculator instructions 1B.2. Generating graphs of functions .... p. 13

Calculator instructions 1B.3. Tips for solving worksheets .... p. 15

REVIEW OF EQUATIONS OF LINES AND CIRCLES

Worksheet 1C.1. Can you match this? .... p. 19

Worksheet 1C.2. Pick a friend .... p. 21

FUNCTIONS

Worksheet 1D.1. Pixels, decimals, coordinates, and angles .... p. 23

Worksheets 1D.2–1D.4. Graphs and values of functions .... p. 25

MAXIMUM/MINIMUM PROBLEMS: A GRAPHICAL APPROACH

Worksheet 1E.1. Minimizing an area .... p. 31

Worksheet 1E.2. Maximizing an area .... p. 33

Worksheet 1E.3. Maximizing the yield from an orchard .... p. 35

Worksheet 1E.4. Diffraction of light .... p. 37

EQUATIONS AND INEQUALITIES WITH POLYNOMIALS

Worksheet 1F.1. Solving equations by factoring .... p. 39

Worksheet 1F.2. Solving equations with the quadratic formula .... p. 41

Worksheet 1F.3. Two equations with absolute values and an inequality .... p. 43

Worksheet 1F.4. Domains of functions involving square roots .... p. 45

LIMITS AND CONTINUITY

Worksheet 1G.1. Two-sided finite limits: Geometric and numerical investigations .... p. 47

Worksheet 1G.2. One-sided finite limits: Geometric and numerical investigations ..... p. 49

Worksheet 1G.3. Infinite limits: Geometric and numerical investigations .... p. 51

Worksheet 1G.4. Continuity and the Extreme and Intermediate value theorems ...... p. 53

---

*This workbook is written for use with a Texas Instruments TI-81 ® calculator. Substitute pages are available for Texas Instruments TI-82 ® and TI-85 ®, Casio fx-7000G ® and fx-7700G ®, and Sharp EL-9200 ® and EL-9300 ® calculators and for the Macintosh ® software Calculus T/L ®.

REVIEW OF TRIGONOMETRIC FUNCTIONS

Worksheet 1H.1. A ferris wheel and the Berlin wall .... p. 55

Worksheet 1H.2. Equations involving trigonometric functions .... p. 57

Worksheet 1H.3. Limits involving trigonometric functions .... p. 59

REVIEW OF LOGARITHMS AND EXPONENTIAL FUNCTIONS

Worksheet 1I.1. Graphs involving logarithms and exponential functions .... p. 61

Worksheet 1I.2. Equations involving logarithms and exponential functions .... p. 63

Worksheet 1I.3. The algebra of logarithms and exponential functions .... p. 65

PART 2. THE DERIVATIVE AND APPLICATIONS

RATES OF CHANGE: INITIAL INVESTIGATIONS

Worksheet 2A.1. Piecewise constant velocity .... p. 67

Worksheet 2A.2. Nonconstant velocity: A graphical approach .... p. 69

Worksheet 2A.3. Other rates of change: A graphical approach .... p. 71

TANGENT LINES AND SECANT LINES: INITIAL INVESTIGATIONS

Secant line program 2B .... p. 73

Worksheet 2C.1. Tangent lines to $y = x^n$: Numerical experiments .... p. 77

Worksheet 2C.2. Tangent lines to $y = \cos x$ and $y = \sin x$: Numerical experiments .... p. 79

GRAPHS AND DIFFERENTIATION FORMULAS

Worksheet 2D.1. Calculus with functions given approximately by graphs .... p. 81

Worksheet 2D.2. Derivatives of composite functions: Initial investigations .... p. 83

Worksheet 2D.3. Derivatives of composite functions: Examples with powers .... p. 85

Worksheet 2D.4. A closer look at some derivatives .... p. 87

Calculator instructions 2E: Graphs of approximate derivatives .... p. 89

RECOGNIZING GRAPHS WITHOUT USING DERIVATIVES

Worksheet 2F.1. Forming polynomials from monomials .... p. 91

Worksheet 2F.2. Factored polynomials and rational functions .... p. 93

Worksheet 2F.3. Fractional and negative powers .... p. 95

USING DERIVATIVES TO STUDY GRAPHS

Worksheet 2G.1. The first-derivative test: Initial investigations .... p. 97

Worksheet 2G.2. The second-derivative test: Initial investigations .... p. 99

Worksheet 2G.3. Graphs of functions constructed from the sine and cosine .... p. 101

Worksheet 2G.4. Graphs with vertical tangent lines, cusps, and limited extent ...... p. 103

RATES OF CHANGE

Worksheet 2H.1. Rates of change of functions whose graphs are lines .... p. 105

Worksheet 2H.2. The rate of change of one distance with respect to another .... p. 107

Worksheet 2H.3. Rates of change of the volume of a tent .... p. 109

Worksheet 2H.4. A rate of change of the height of a ladder .... p. 111

Worksheet 2H.5. A rate of change of the length of a shadow .... p. 113

Graphing calculator workbook — iii

NEWTON'S METHOD

    Newton's method program 2I .... p. 115

    Worksheet 2J.1. Newton's method: Finding the dimensions of a cone .... p. 121

    Worksheet 2J.2. Newton's method: Minimizing the light from two lamps .... p. 123

    Worksheet 2J.3. Newton's method: Studying the graph of a function .... p. 125

    Worksheet 2J.4. Newton's method: Finding the closest point to a curve .... p. 127

PART 3. THE INTEGRAL AND APPLICATIONS

AREAS IN APPLICATIONS

    Worksheet 3A.1. Finding distances from velocities: Initial investigations .... p. 129

    Worksheet 3A.2. Volumes by slicing: Initial investigations .... p. 131

    Worksheet 3A.3. Lengths of curves and average values: Initial investigations .... p. 133

RIEMANN SUMS AND RIEMANN INTEGRALS

    Riemann sum program 3B .... p. 135

INTEGRATION FORMULAS

    Worksheet 3C.1. Integrals of x^n: Numerical experiments .... p. 141

    Worksheet 3C.2. Integrals of $\cos x$ and $\sin x$: Numerical experiments .... p. 143

    Worksheet 3C.3. Algebra of integrals: Numerical experiments .... p. 145

    Worksheet 3D.1. The Fundamental theorem: A geometric interpretation .... p. 147

    Worksheet 3D.2. The Fundamental theorem: Examples and nonexamples .... p. 149

SIMPSON'S RULE

    Simpson's rule program 3E .... p. 151

THREE-DIMENSIONAL MODELS

    Construction 3F.1. Volumes by slicing .... p. 157

    Construction 3F.2. Solids of revolution .... p. 161

PROJECTS

    Project 3G.1. Designing a shish-kabob .... p. 167

    Project 3G.2. Billy the Kid .... p. 169

    Project 3G.3. Designing a boat .... p. 171

PART 4. OTHER TRANSCENDENTAL FUNCTIONS

NUMERICAL EXPERIMENTS

    Worksheet 4A.1. Derivatives of exponential functions: Numerical experiments .... p. 173

    Worksheet 4A.2. Derivatives of logarithms: Numerical experiments .... p. 175

    Worksheet 4A.3. The curious integrals of $1/x$: Numerical experiments .... p. 177

RECOGNIZING GRAPHS IN APPLICATIONS

    Worksheet 4B.1. Applications of exponential functions .... p. 179

    Worksheet 4B.2. Applications of logarithms .... p. 181

COMPOUND INTEREST

    Worksheet 4B.3. Compound interest .... p. 183

MAXIMUM/MINIMUM PROBLEMS

    Worksheet 4C.1. Masterpiece around the corner .... p. 185

    Worksheet 4C.2. What's the angle? .... p. 187

INVERSE TRIGONOMETRIC FUNCTIONS

    Worksheet 4D.1. Functions constructed from inverse trigonometric functions .... p. 189

PARTIAL FRACTIONS

    Worksheet 4E.1. Integrating rational functions: Initial investigations .... p. 191

PART 5. SEPARABLE DIFFERENTIAL EQUATIONS

INITIAL INVESTIGATIONS

    Worksheet 5A.1. Differential equations of growth and decay .... p. 193

    Worksheet 5A.2. Differential equations of velocity with resistance .... p. 195

    Worksheet 5A.3. Differential equations of families of curves .... p. 197

    Direction field program 5B .... p. 199

NUMERICAL METHODS

    Worksheet 5C.1. Euler's method .... p. 207

    Runge-Kutta program 5D .... p. 209

PART 6. INFINITE SEQUENCES AND SERIES

GEOMETRIC SERIES: INITIAL INVESTIGATIONS

    Worksheet 6A.1. Going part way .... p. 217

    Worksheet 6A.2. Loan payments and St. Ives .... p. 219

    Worksheet 6A.3. A flip of a coin, a roll of the dice .... p. 221

INFINITE SEQUENCES

    Sequence program 6B .... p. 223

    Worksheet 6C.1. Infinite sequences: Initial investigations .... p. 227

    Worksheet 6C.2. The ϵN- and YN-definitions of convergence and divergence to ∞ .. p. 229

INFINITE SERIES OF CONSTANTS

    Series program 6D. .... p. 231

    Worksheet 6E.1. The Comparison test for series with nonnegative terms .... p. 237

    Worksheet 6E.2. Alternating series .... p. 239

    Worksheet 6E.3. Using convergence and divergence tests .... p. 241

TAYLOR POLYNOMIALS AND POWER SERIES

    Worksheet 6F.1. Matching functions to Taylor polynomial approximations .... p. 243

    Worksheet 6F.2. Adding more terms; Looking near and far away .... p. 245

    Worksheet 6F.3. Error estimates vs actual errors .... p. 247

    Worksheet 6F.4. Operations with power series .... p. 249

Index .... p. 251

PREFACE

This *Workbook* contains exploratory worksheets and other assignments for a graphing calculator or computer based course in one-variable calculus. It can be used with any calculus text and any graphing calculator or computer with graphing software.

Teaching philosophy

The EXPLORATORY WORKSHEETS introduce new topics with exercises that require only precalculus techniques or basic calculus facts that have been covered earlier in the course. The goal of these assignments is to have students investigate problems on their own, using a calculator or computer to generate graphs and carry out numerical experiments, before they study the related calculus results and procedures. The exploratory worksheets give students a preliminary understanding of basic concepts, so they are more motivated and better prepared to learn from lectures and textbooks, and so they realize that calculus is more than a fixed catalog of types of examination questions.

Other WORKSHEETS train students to use a calculator or computer in solving problems and for checking answers and give them experience dealing with instances when calculators or computers give incorrect results. Key-by-key instructions are provided for calculator programs with graphics for secant line approximations of tangent lines, Newton's method, Riemann sums, Simpson's rule, a Runge-Kutta method for finding approximate solutions of differential equations, and for plotting direction fields, values of sequences, and partial sums of infinite series.

Part 3 on integration includes PROJECTS, which require more input from students than most assignments, and three-dimensional CONSTRUCTIONS, which help students understand the method of slicing and visualize solids of revolution.

Flexibility

Outlines for regular and short calculus classes describe how the workbook can be integrated with traditional material from any calculus text. Worksheets and other assignments that are listed in the outlines before the descriptions of lectures and reading assignments on the same topics can be used to prepare students for the lectures.

> †Assignments marked with daggers in the outlines should be partially completed in supervised workshop sessions where students can ask questions about the problems.

> ‡Assignments marked with double daggers are suitable for unsupervised homework.

Instructions and programs for Texas Instruments TI-81 calculators are included. Substitute pages for Texas Instruments TI-82 and TI-85, Casio fx-7000G and fx-7700G, and Sharp EL-9200 and EL-9300 calculators and for the Macintosh software Calculus T/L are available. The instructions and programs can be modified for other calculators and computer packages.

Implementation

In planning your class, you may need to omit some topics and types of exercises which you have included in the past in order to have time for the workbook. Exercises in your text that illustrate concepts can be skipped if those concepts are adequately covered in the worksheets. You might also omit some applications and special procedures—such as finding limits by rationalizing differences of square roots and some of the more involved techniques of integration—which students will not need in subsequent courses or which they could learn later if needed.

Some suggestions:

- Make it clear to your class that you have planned your course carefully and that you believe it will increase their ability to learn mathematics.

- Have your students do workbook assignments in groups of three or four in and outside of class. Use part of the first class meeting as a workshop session with temporary group assignments. At that meeting have them fill out a questionnaire listing the precalculus and calculus classes they have taken and the grades they earned, their local addresses, the names of any students with whom they would like to work, and the times they would be free to meet outside of class. Use this information to assign groups that facilitate their working together and so that no group consists solely of weak students. Require that groups give you minutes of their meetings to help you monitor and encourage their activities. Also, talk to other instructors who have used collaborative learning techniques so you can benefit from their experience.

- Hold as many workshop sessions as you can schedule, especially at the beginning of a course—to help students learn to use their calculators or computers and to get them accustomed to working together—and later whenever exploratory assignments are used to introduce new topics. If you have a large class, find a room for workshops with at least half again as many seats as students and have students sit with every third row empty so you can reach all of them. Use graduate or undergraduate tutors to help you answer questions. Prepare tutors each day with brief explanations of the types of questions they might be asked and how to answer them.

- If you assign a worksheet in a supervised workshop session, give students as few verbal instructions as possible and do not show them similar solved examples. Let them get as far as they can on their own, and then give groups or individuals whatever guidance they need. Have them finish the assignments outside of class. If you give a worksheet as unsupervised homework, be sure the class knows in advance what is expected on it.

- Give students plenty of time to complete workbook problems, and require that they describe their results in clear and correct English.

- Make workbook activities a significant part of the course, have students' work from it scored carefully, and discuss the solutions at least briefly in class.

- When possible, have the exploratory worksheets on a new topic be due at the beginning of the class when you first discuss it, and use problems from the worksheets as examples in your lecture. You can then explain what was expected on the worksheets while you develop the underlying ideas with calculus.

- Write examinations that give students the opportunity to display their understanding of concepts and their ability to use a calculator or computer. Use the imperative "explain" frequently in your questions.

One, unexpected, change in my classes with the use of exploratory assignments and graphing calculators has been the improved atmosphere in my large lectures. Students ask more and better questions and are more involved than they were in my years of giving traditional lectures. I believe this is because they actively particpate with their graphing calculators in class; because I relate personally to individuals during in-class workshop sessions; because the students are more accustomed to discussing their assignments together; and because I am asking them to think more and memorize less.

Class-room testing

Preliminary versions of this workbook have been used since the summer of 1991 in large and small classes at UCSD, the University of New Hampshire, Georgia Institute of Technology, the University of Nevada, Las Vegas, and other colleges and high schools. The students in these courses generally liked using calculators and had no difficulty learning how to use them. Some students and teaching assistants resisted the changes in teaching methods—such as the requiring of group work, the de-emphasis of learning by studying solved examples, and the emphasis on giving verbal descriptions of concepts and problem solutions—but this resistance faded as the courses progressed.

Supplements

Three supplements are available to adopters of the workbook:

- The substitute pages for various types of calculators and Calculus T/L that are described above.

- An instructor's manual, titled *Graphs and Answers*, with descriptions of all workbook assignments and suggestions for their use, all answers, and computer-generated versions of all graphs that students are required to generate on their calculators or computers in solving the problems.

- A students' study guide for classes that use Volume I of *Calculus and Analytic Geometry* by Al Shenk (Fourth edition, Harper Collins, 1988). This guide lists reading assignments and exercises from the text that complement the workbook, with descriptions of the material from both sources and study suggestions. Topics from the text that some instructors are likely to skip are labeled as "review" or "elective" material to help instructors plan their classes and assign homework.

Acknowledgments

I am grateful to George Duda of Harper Collins College Publishing Company, Bill Helton of the University of California at San Diego, Phyllis Hillis of Oak Ridge High School, and all the others who have contributed to this project and supported me in working on it, including Jamal Bernhard, University of California, San Diego (UCSD); Victor Cifarelli, UCSD; Mahmoud Fath El-Den, Fort Hays State University; Joan Ferrini-Mundy, University of New Hampshire; Jay Fillmore, UCSD; Skip Garibaldi, UCSD; Karen Graham, University of New Hampshire; LaDawn Haws, California State University, Chico; Ziv Hellman, UCSD; Theresa Jones, Amarillo College; Rose Kaplan, Ohio State University, Newark; John Neff, Georgia Institute of Technology; Dick Pilgrim, UCSD; Raymond Rolwing, University of Cincinnati; Beth Smith, Grossmont Community College; Tom Schaffter, Fort Lewis College; Gerald White, Western Illinois University,;Jiangang Ying, UCSD; and Theodore Wilcox, Rochester Institute of Technology.

Feel free to contact me for any further information. Also, I would like to hear about your experience with new teaching techniques: what you have tried, what works, what does not, and what you will try next time.

Al Shenk, Math Department
UCSD, La Jolla, CA 92093
email: ashenk@ucsd.edu

Graphing calculator workbook ix

TO THE STUDENT

The worksheets in this manual have three main objectives:

- To show you how a graphing calculator or computer can be used to help clarify concepts, solve problems, and check results.

- To give you more opportunities to work problems on your own, before you study explanations and solved examples from lectures or a textbook.

- To give you practice discussing and writing mathematics.

Learning calculus

The basic ideas of calculus are not difficult but may not be very meaningful until you have worked with them over a period of time. Your understanding will develop and change with experience. On the other hand, to obtain the most benefit from your efforts, you have to do more than just sit through lectures and do your homework. You need to think about what you are doing so you can see from your own point of view how and why calculus works the way it does. Some suggestions:

- Look for the visual meanings of definitions, results, and procedures with graphs of functions and other drawings.

- Experiment with specific values of variables in algebraic calculations—with and without a calculator or computer—to see underlying patterns, to get a general idea of what answers to expect, and to check your results.

- Clarify your understanding by expressing your ideas in words. Whenever you figure out some aspect of calculus, such as the purpose of a definition, the meaning of a result, or the line of reasoning in a proof or problem solution, describe it to yourself or a fellow student in everyday English.

Using the worksheets

Some of the worksheets introduce new topics and ideas by using only precalculus techniques or relatively easy calculus ideas that you have studied earlier. Work these without consulting your textbook. The understanding you gain on your own will help you learn from your book later when you do read it. Other worksheets use techniques discussed in your book.

Allow plenty of time for each worksheet. Even though your instructor uses only a few minutes to work an example or derive a result in class, that is not how mathematics is done. It generally takes at least a couple of days to solve a new type of problem and much longer to to obtain a comprehensive understanding of a significant new idea.

Your efforts on each worksheet should be in several stages:

- First, follow the instructions on the worksheet, putting your calculations, graphs, and any tentative conclusions on scratch paper. Buy a lined, transparent, plastic ruler for drawing and reading graphs.

- If you have trouble with a problem, set it aside and think about your results and the purpose of the assignment in spare moments before continuing with it another day.

- Use graphs and numerical calculations, with the programs in the workbook as appropriate, to check your answers. Bear in mind, however, that graphs and approximate calculations sometimes give significantly incorrect results.

- Compare your work and discuss your interpretations with at least one other student in your class.

- If you have any questions about the topic that you and your fellow students cannot answer or if you have trouble carrying out any instructions in the worksheet, ask your professor or a teaching assistant for help.

- Once you feel you understand the worksheet, repeat your calculations to check your results. Then state your conclusions in a few sentences, rewriting them as necessary to express yourself clearly. Put the final draft of your calculations, graphs, and conclusions on the worksheet or on extra sheets of clean paper—in pencil so you can make corrections.

Also apply these principles, as appropriate, to the rest of your work in the course. Make it your goal to understand ideas and results in your own terms and from your own point of view and develop procedures for solving problems that work best for you. Whenever possible, read ahead in the textbook so you have some familiarity with topics to be covered in lectures. You can learn much more from a lecture if you can follow what the instructor is saying and relate it to your own ways of thinking about the subject.

Calculators and computers

The worksheets require either a graphing calculator or a computer with graphing software. Instructions are included for Texas Instruments TI-81 calculators, along with programs for performing operations that are not provided as built-in features. Substitute pages are available through HarperCollinsCollege representatives for Texas Instruments TI-82 and TI-85, Casio fx-7000G and fx-7700G, and Sharp EL-9200 and EL-9300 calculators, and for the Macintosh software Calculus T/L. If your class uses other types of calculators or computer packages, your instructor will provide information on how to use them.

Graphing calculator workbook

COURSE OUTLINE I, FOR ENGINEERING, SCIENCE, AND MATHEMATICS MAJORS

PART 1. FUNCTIONS, GRAPHS, AND LIMITS

INTRODUCTION

WORKBOOK

Worksheets 1A.1–1A.4. Maximum/minimum problems: Initial investigations[†]

LECTURES AND TEXTBOOK

Overview of the subject and the course

How calculators work and how they deceive

FUNCTIONS AND THEIR GRAPHS

LECTURES AND TEXTBOOK

Functions: Notation, domains, and graphs

Examples of functions from applications

WORKBOOK

Calculator instructions 1B.1 and 1B.2. Basic operations and generating graphs[‡]

Worksheet 1D.1. Pixels, decimals, coordinates, and angles[†]

Worksheets 1E.1–1E.4. Maximum/minimum problems: A graphical approach[†]

LIMITS AND CONTINUITY

LECTURES AND TEXTBOOK

The geometric and numerical meaning of finite limits

The geometric and numerical meaning of continuity

Limit theorems

Finding limits of rational functions by factoring

Infinite limits

The Extreme and Intermediate value theorems

WORKBOOK

Worksheets 1G.1–1.G.3. Finite and infinite limits: Geometric and numerical investigations[‡]

Worksheet 1G.4. Continuity and the Extreme and Intermediate value theorems[†]

OTHER PART 1 WORKBOOK ASSIGNMENTS

MORE ON ANALYTIC GEOMETRY

Worksheet 1C.1. Can you match this? (Equations of lines)[‡]

Worksheet 1C.2. Pick a friend (Equations of circles and semi-circles)[‡]

ALGEBRA REVIEW AND CALCULATOR PRACTICE

Worksheets 1D.2–1D.4. Graphs and values of functions[†]

Worksheets 1F.1–1F.4. Equations and inequalities with polynomials[‡]

REVIEW OF TRIGONOMETRIC FUNCTIONS

Worksheet 1H.1. A ferris wheel and the Berlin wall[†]

Worksheet 1H.2. Equations involving trigonometric functions[‡]

Worksheet 1H.3. Limits involving trigonometric functions[‡]

REVIEW OF LOGARITHMS AND EXPONENTIAL FUNCTIONS

Worksheet 1I.1. Graphs involving logarithms and exponential functions[†]

Worksheet 1I.2. Equations involving logarithms and exponential functions[†]

Worksheet 1I.3. The algebra of logarithms and exponential functions[‡]

[†]Written to be at least partially completed in supervised workshop sessions.

[‡]Suitable for homework or workshop sessions

Part 2. THE DERIVATIVE AND APPLICATIONS

INTRODUCTION

WORKBOOK

 Worksheet 2A.1. Piecewise constant velocity[†]
 Worksheet 2A.2. Nonconstant velocity: A graphical approach[†]
 Worksheet 2A.3. Other rates of change: A graphical approach[†]

LECTURES AND TEXTBOOK

 Overview of the derivative: History and applications

DEFINITION OF THE DERIVATIVE

LECTURES AND TEXTBOOK

 Finding derivatives by the definition
 Average and instantaneous rates of change

FINDING DERIVATIVES

WORKBOOK

 Secant line program 2B[‡]
 Worksheet 2C.1. Tangent lines to $y = x^n$: Numerical experiments[‡]
 Worksheet 2C.2. Tangent lines to $y = \sin x$ and $y = \cos x$: Numerical experiments[‡]

LECTURES AND TEXTBOOK

 Derivatives of x^n and of linear combinations
 The product and quotient rules
 Derivatives of trigonometric functions

WORKBOOK

 Worksheet 2D.1. Calculus with functions given approximately by graphs[†]
 Worksheet 2D.2. Derivatives of composite functions: Initial investigations[†]

LECTURES AND TEXTBOOK

 The chain rule

WORKBOOK

 Worksheet 2D.3. Derivatives of composite functions: Examples with powers[‡]
 Calculator instructions 2E: Graphs of approximate derivatives[‡]

LECTURES AND TEXTBOOK

 Implicit differentiation

SKETCHING GRAPHS WITHOUT A CALCULATOR OR COMPUTER

WORKBOOK

 Worksheets 2F.1–2F.3. Recognizing graphs without using derivatives[†]
 Worksheet 2G.1. The first-derivative test: Initial investigations[‡]
 Worksheet 2G.2. The second-derivative test: Initial investigations[‡]

LECTURES AND TEXTBOOK

 The mean value theorem
 Sketching raphs of rational functions by plotting points and using derivative tests

ANALYZING CALCULATOR OR COMPUTER GENERATED GRAPHS

LECTURES AND TEXTBOOK

 Periodicity, domains, asymptotes, vertical tangent lines, corners, and cusps

WORKBOOK

 Worksheet 2G.3. Graphs of functions constructed from the sine and cosine[‡]
 Worksheet 2G.4. Graphs with vertical tangent lines, cusps, and limited extent[‡]

[†]Written to be at least partially completed in supervised workshop sessions.
[‡]Suitable for homework or workshop sessions

Graphing calculator workbook
Course outline I

APPLICATIONS
LECTURES AND TEXTBOOK
    Maximum/minimum problems

WORKBOOK
    Worksheets 2H.1–2H.5. Rates of change[†]

LECTURES AND TEXTBOOK
    Related rate problems

OTHER PART 2 WORKBOOK ASSIGNMENTS
MORE ON DERIVATIVES
    Worksheet 2D.4. A closer look at some derivatives[†]
NEWTON'S METHOD FOR FINDING APPROXIMATE SOLUTIONS OF $F(x) = 0$
    Newton's method program 2I[‡]
    Worksheets 2J.1–2J.4. Using Newton's method[‡]

PART 3. THE INTEGRAL AND APPLICATIONS

INTRODUCTION
WORKBOOK
    Worksheets 3A.1–3A.3. Areas in applications[†]

LECTURES AND TEXTBOOKS
    Overview of the integral: History and applications

THE DEFINITION OF THE INTEGRAL
LECTURES AND TEXTBOOK
    Riemann sums and Riemann integrals

WORKBOOK
    Riemann sum program 3B[‡]

FINDING INTEGRALS
    Worksheets 3C.1 and 3C.2. Integrals of x^n, $\cos x$, and $\sin x$: Numerical experiments[†]
    Worksheet 3C.3. Algebra of integrals: Numerical experiments[†]

LECTURES AND TEXTBOOK
    The Fundamental theorem
    Indefinite integrals
    Algebra of integrals
    Integrals of x^n $(n \neq -1)$, $\cos x$, $\sin x$, and other trigonometric functions

WORKBOOK
    Worksheet 3D.1. The Fundamental theorem: A geometric interpretation[†]
    Worksheet 3D.2. The Fundamental theorem: Examples and nonexamples[†]

LECTURES AND TEXTBOOK
    Integration by substitution

APPLICATIONS (A SELECTION)
LECTURES AND TEXTBOOK
    Areas between curves
    Volumes of solids of revolution
    Volumes by slicing
    Integrals of rates of change

[†]Written to be at least partially completed in supervised workshop sessions.

[‡]Suitable for homework or workshop sessions

Work and energy
Average values
Lengths of graphs of functions
Centers of gravity and centroids; Pappus's theorems

WORKBOOK
Construction 3F.1. Volumes by slicing[‡]
Construction 3F.2. Solids of revolution[‡]
Project 3G.1. Designing a shish kabob[‡]
Project 3G.2. Billy the Kid[‡]
Project 3G.3. Designing a boat[‡]

OTHER PART 3 WORKBOOK ASSIGNMENTS
SIMPSON'S RULE
Simpson's rule program 3E[‡]

PART 4. OTHER FUNCTIONS AND INTEGRATION TECHNIQUES

INTRODUCTION
LECTURES AND TEXTBOOK
Overview: Logarithms and exponential functions. History and applications.

LOGARITHMS AND EXPONENTIAL FUNCTIONS
WORKBOOK
Worksheets 4A.1 and 4A.2. Derivatives of logarithms and exponential functions: Numerical experiments[†]
Worksheet 4A.3. The curious integrals of $1/x$: Numerical experiments[†]

LECTURES AND TEXTBOOK
Inverse functions
$\ln x$ and e^x
$\log_b x$ and b^x
The differential equation $dy/dt = ry$
Compound interest

WORKBOOK
Worksheet 4B.1. Applications of exponential functions[‡]
Worksheet 4B.2. Applications of logarithms[†]
Worksheet 4B.3. Compound interest[‡]

INVERSE TRIGONOMETRIC AND HYPERBOLIC FUNCTIONS
LECTURES AND TEXTBOOK
The inverse trigonometric functions
Hyperbolic functions

WORKBOOK
Worksheet 4C.1: Masterpiece around the corner[‡]
Worksheet 4C.2. What's the angle?[‡]
Worksheet 4D.1. Functions constructed from inverse trigonometric functions[†]

L'HOPITAL'S RULE
LECTURES AND TEXTBOOK
L'Hopital's rule

[†]Written to be at least partially completed in supervised workshop sessions.
[‡]Suitable for homework or workshop sessions

Graphing calculator workbook
Course outline I

FURTHER TECHNIQUES OF INTEGRATION

LECTURES AND TEXTBOOK
Integration by parts
Integration using trigonometric identities
Integration by inverse trigonometric substitutions

WORKBOOK
Worksheet 4E.1. Integrating rational functions: Initial investigations[†]

LECTURES AND TEXTBOOKS
Integration by partial fractions
Improper integrals

PART 5. SEPARABLE DIFFERENTIAL EQUATIONS

INTRODUCTION

LECTURES AND TEXTBOOK
Differential equations: History and applications
Overview: First-order differential equations and direction fields

WORKBOOK
Worksheet 5A.1. Differential equations of growth and decay[†]
Worksheet 5A.2. Differential equations of velocity with resistance[†]
Worksheet 5A.3. Differential equations of families of curves[†]
Direction field program 5B[‡]

TECHNIQUES

LECTURES AND TEXTBOOK
Separable differential equations
Applications

NUMERICAL METHODS

WORKBOOK
Worksheet 5C.1. Euler's method[‡]

LECTURES AND TEXTBOOK
Euler's method and improvements on it

WORKBOOK
Runge-Kutta program 5D[‡]

PART 6. INFINITE SEQUENCES AND SERIES

INTRODUCTION

WORKBOOK
Worksheets 6A.1–6A.3. Geometric series: Initial investigations[†]
Sequence program 6B[‡]
Worksheet 6C.1. Infinite sequences: Initial experiments[‡]

INFINITE SEQUENCES

LECTURES AND TEXTBOOK
Convergence and divergence of infinite sequences
The ϵN- and YN-definitions of convergence and divergence to $\pm\infty$

WORKBOOK
Worksheet 6C.2. The ϵN- and YN-definitions of convergence and divergence to ∞[†]

[†]Written to be at least partially completed in supervised workshop sessions.
[‡]Suitable for homework or workshop sessions

xvi Graphing calculator workbook
Course outline I

INFINITE SERIES OF CONSTANTS

WORKBOOK

Series program 6D‡

LECTURES AND TEXTBOOK

Convergence and divergence of infinite series
The Comparison and Limit comparison tests

WORKBOOK

Worksheet 6E.1. The Comparison test for series with nonnegative terms‡
Worksheet 6E.2. Alternating series‡

LECTURES AND TEXTBOOK

Absolute convergence and alternating series
The Integral and Ratio tests

WORKBOOK

Worksheet 6E.3. Using convergence and divergence tests‡

TAYLOR POLYNOMIALS AND POWER SERIES

WORKBOOK

Workshet 6F.1. Matching functions to Taylor polynomial approximations[†]

LECTURES AND TEXTBOOK

Taylor polynomials and Taylor's theorem

WORKBOOK

Worksheet 6F.2. Adding more terms; Looking near and far away[†]
Worksheet 6F.3. Error estimates vs. actual errors[†]

LECTURES AND TEXTBOOK

Taylor and power series
Intervals (or radii) of convergence
Algebra and calculus of power series

WORKBOOK

Worksheet 6F.4. Operations with power series‡

[†]Written to be at least partially completed in supervised workshop sessions.
[‡]Suitable for homework or workshop sessions

COURSE OUTLINE II, FOR BIOLOGY, SOCIAL SCIENCE, AND HUMANITIES MAJORS

PART 1. FUNCTIONS, GRAPHS, AND LIMITS

INTRODUCTION

WORKBOOK

Worksheets 1A.1–1A.4. Maximum/minimum problems: Initial investigations[†]

LECTURES AND TEXTBOOK

Overview of the subject and the course

How calculators work and how they deceive

Review of coordinates, the distance formula, and equations of lines

WORKBOOK

Calculator instructions 1B.1–1B.2. Basic operations and generating graphs[‡]

Worksheet 1C.1. Can you match this? (Equations of lines)[†]

Worksheet 1C.2. Pick a friend (Equations of circles and semi-circles)[‡]

FUNCTIONS AND THEIR GRAPHS

LECTURES AND TEXTBOOK

Functions: Notation, domains, and graphs

Functions given approximately by their graphs

Linear and quadratic functions

Review of factoring, completing the square, and the quadratic formula

WORKBOOK

Worksheet 1D.1. Pixels, decimals, coordinates, and angles[†]

Worksheets 1D.2–1D.4. Graphs and values of functions[†]

Worksheets 1E.1–1E.4. Maximum/minimum problems: A graphical approach[†]

Worksheets 1F.1–1F.2. Factoring and the quadratic formula[‡]

LIMITS AND CONTINUITY

LECTURES AND TEXTBOOK

The geometric and numerical meaning of finite limits

The geometric and numerical meaning of continuity

Limit theorems

Finding limits of rational functions by factoring

Infinite limits

WORKBOOK

Worksheets 1G.1–1.G.3. Finite and infinite limits: Geometric and numerical investigations[‡]

OTHER PART 1 WORKBOOK ASSIGNMENTS

MORE ON POLYNOMIALS

Worksheet 1F.3. Two equations with absolute values and an inequality[‡]

Worksheet 1F.4. Domains of functions involving square roots[‡]

MORE ON LIMITS AND CONTINUITY

Worksheet 1G.4. Continuity and the Extreme and Intermediate value theorems[†]

REVIEW OF TRIGONOMETRIC FUNCTIONS

Worksheet 1H.1. A ferris wheel and the Berlin wall[†]

Worksheet 1H.2. Equations involving trigonometric functions[‡]

Worksheet 1H.3. Limits involving trigonometric functions[‡]

[†]Written to be at least partially completed in supervised workshop sessions.

[‡]Suitable for unsupervised homework.

REVIEW OF LOGARITHMS AND EXPONENTIAL FUNCTIONS
>   Worksheet 1I.1. Graphs involving logarithms and exponential functions[†]
>   Worksheet 1I.2. Equations involving logarithms and exponential functions[‡]
>   Worksheet 1I.3. The algebra of logarithms and exponential functions[‡]

PART 2. THE DERIVATIVE AND APPLICATIONS

INTRODUCTION
WORKBOOK
>   Worksheet 2A.1. Piecewise constant velocity[†]
>   Worksheet 2A.2. Nonconstant velocity: A graphical approach[†]
>   Worksheet 2A.3. Other rates of change: A graphical approach[†]

LECTURES AND TEXTBOOK
>   Overview of the derivative: History and applications

DEFINTION OF THE DERIVATIVE
LECTURES AND TEXTBOOK
>   Finding derivatives by the definition
>   Average and instantaneous rates of change

FINDING DERIVATIVES
WORKBOOK
>   Secant line program 2B[‡]
>   Worksheet 2C.1. Tangent lines to $y = x^n$: Numerical experiments[†]
>   Worksheet 2C.2. Tangent lines to $y = \cos x$ and $y = \sin x$: Numerical experiments[†]

LECTURES AND TEXTBOOK
>   Derivatives of x^n and of linear combinations
>   The product and quotient rules
>   Derivatives of trigonometric functions

WORKBOOK
>   . Worksheet 2D.1. Calculus with functions given approximately by graphs[†]
>   Worksheet 2D.2. Derivatives of composite functions: Initial investigations[†]

LECTURES AND TEXTBOOK
>   The chain rule

WORKBOOK
>   Calculator instructions 2E: Graphs of approximate derivatives[‡]

LECTURES AND TEXTBOOK
>   Implicit differentiation

SKETCHING GRAPHS WITHOUT A CALCULATOR OR COMPUTER
WORKBOOK
>   Worksheets 2F.1–2F.3. Recognizing graphs without using derivatives[†]
>   Worksheets 2G.1. The first-derivative test: Initial investigations[‡]
>   Worksheets 2G.2. The second-derivative test: Initial investigations[‡]

LECTURES AND TEXTBOOK
>   Sketching graphs of polynomials by plotting points and using derivative tests

---

[†]Written to be at least partially completed in supervised workshop sessions.
[‡]Suitable for unsupervised homework.

Graphing calculator workbook
Course outline II

ANALYZING CALCULATOR OR COMPUTER GENERATED GRAPHS

LECTURES AND TEXTBOOK
Periodicity, domains, asymptotes, vertical tangent lines, and cusps

WORKBOOK
Worksheet 2G.3. Graphs of functions constructed from the sine and cosine[†]
Worksheet 2G.4. Graphs with vertical tangent lines, cusps, and limited extent[‡]

APPLICATIONS

LECTURES AND TEXTBOOK
Maximum/minimum problems

WORKBOOK
Worksheets 2H.1–2H.6. Rates of change[†]

LECTURES AND TEXTBOOK
Related rate problems

OTHER PART 2 WORKBOOK ASSIGNMENTS

MORE ON COMPOSITE FUNCTIONS
Worksheet 2D.3. Derivatives of composite functions: Examples with powers[†]

MORE ON THE DEFINITION OF DERIVATIVES
Worksheet 2D.4. A closer look at some derivatives[†]

NEWTON'S METHOD FOR FINDING APPROXIMATE SOLUTIONS OF $F(x) = 0$
Newton's method program 2I[‡]
Worksheets 2J.1–2J.4. Using Newton's method[‡]

PART 3. THE INTEGRAL AND APPLICATIONS

INTRODUCTION

WORKBOOK
Worksheets 3A.1–3A.3. Areas in applications[†]

LECTURES AND TEXTBOOKS
Overview of the integral: History and applications

THE DEFINITION OF THE INTEGRAL

LECTURES AND TEXTBOOK
Riemann sums and Riemann integrals

WORKBOOK
Riemann sum program 3B[‡]

FINDING INTEGRALS

WORKBOOK
Worksheet 3C.1. Integrals of x^n: Numerical experiments[†]
Worksheet 3C.2. Integrals of $\cos x$ and $\sin x$: Numerical experiments[†]
Worksheet 3C.3. Algebra of integrals: Numerical experiments[†]

LECTURES AND TEXTBOOK
The Fundamental theorem
Indefinite integrals
Integrals of x^n $(n \neq -1)$, $\cos x$, $\sin x$ and other trigonometric functions
Integration by substitution

[†]Written to be at least partially completed in supervised workshop sessions.
[‡]Suitable for homework or workshop sessions

APPLICATIONS (A SELECTION)

LECTURES AND TEXTBOOK
Areas between curves
Volumes of solids of revolution
Volumes by slicing
Integrals of rates of change
Average values
Lengths of graphs of functions

WORKBOOK
Construction 3F.1. Volumes by slicing[‡]
Construction 3F.2. Solids of revolution[‡]

OTHER PART 3 WORKBOOK ASSIGNMENTS

THE FUNDAMENTAL THEOREM
Worksheet 3D.1. The Fundamental theorem: A geometric interpretation[†]
Worksheet 3D.2. The Fundamental theorem: Examples and nonexamples[†]

SIMPSON'S RULE
Simpson's rule program 3E[‡]

PROJECTS
Project 3G.1. Designing a shish-kabob [‡]
Project 3G.2. Billy the Kid[‡]
Project 3G.3. Designing a boat[‡]

PART 4. OTHER FUNCTIONS AND INTEGRATION TECHNIQUES

INTRODUCTION

LECTURES AND TEXTBOOK
Overview: Logarithms and exponential functions. History and applications.

LOGARITHMS AND EXPONENTIAL FUNCTIONS

WORKBOOK
Worksheet 4A.1 and 4A.2. Derivatives of logarithms and exponential functions: Numerical experiments[†]
Worksheet 4A.3. The curious integrals of $1/x$[†]

LECTURES AND TEXTBOOK
$\ln x$ and e^x
$\log_b x$ and b^x
The differential equation $dy/dt = ry$
Compound interest

WORKBOOK
Worksheet 4B.1. Applications of exponential functions[†]
Worksheet 4B.2. Applications of logarithms[†]
Worksheet 4B.3. Compound interest[‡]

FURTHER TECHNIQUES OF INTEGRATION

LECTURES AND TEXTBOOK
Integration by parts

[†]Written to be at least partially completed in supervised workshop sessions.
[‡]Suitable for homework or workshop sessions

Graphing calculator workbook
Course outline II

OTHER PART 4 WORKBOOK ASSIGNMENTS

MAXIMUM/MINIMUM PROBLEMS
Worksheet 4C.1. Masterpiece around the corner.[‡]
Worksheet 4C.2. What's the angle?[‡]

INVERSE TRIGONOMETRIC FUNCTIONS
Worksheet 4D.1. Functions constructed from inverse trigonometric functions[‡]

INTEGRATION BY PARTIAL FRACTIONS
Worksheet 4E.1. Integrating rational functions: Initial investigations[†]

PART 5. OTHER DIFFERENTIAL EQUATIONS

INTRODUCTION

LECTURES AND TEXTBOOK
Differential equations: History and applications
First-order differential equations and direction fields

WORKBOOK
Worksheet 5A.1. Differential equations of growth and decay[†]
Worksheet 5A.2. Differential equations of velocity with resistance[†]
Worksheet 5A.3. Differential equations of families of curves[†]
Direction field program 5B[‡]

TECHNIQUES

LECTURES AND TEXTBOOK
Separable differential equations
Applications

OTHER PART 5 WORKBOOK ASSIGNMENTS

NUMERICAL METHODS
Worksheet 5C.1. Euler's method[‡]
Runge-Kutta program 5D[‡]

PART 6. INFINITE SEQUENCES AND SERIES

INTRODUCTION

WORKBOOK
Worksheets 6A.1–6A.3. Geometric series: Initial investigations[†]

INFINITE SEQUENCES

WORKBOOK
Sequence program 6B[‡]
Worksheet 6C.1. Infinite sequences: Initial experiments[†]

LECTURES AND TEXTBOOK
Convergence and divergence of infinite sequences

INFINITE SERIES OF CONSTANTS

WORKBOOK
Series program 6D[‡]

LECTURES AND TEXTBOOK
Convergence and divergence of infinite series
The Comparison test

[†]Written to be at least partially completed in supervised workshop sessions.
[‡]Suitable for homework or workshop sessions

WORKBOOK
    Worksheet 6E.1. The Comparison test for series with nonnegative terms[‡]
    Worksheet 6E.2. Alternating series[‡]

LECTURES AND TEXTBOOK
    Absolute convergence and alternating series
    The Ratio test

WORKBOOK
    Worksheet 6E.3. Using convergence and divergence tests[‡]

TAYLOR POLYNOMIALS AND POWER SERIES

WORKBOOK
    Workshet 6F.1. Matching functions to Taylor polynomial approximations[†]

LECTURES AND TEXTBOOK
    Taylor polynomials

WORKBOOK
    Worksheet 6F.2. Adding more terms; Looking near and far away[†]

LECTURES AND TEXTBOOK
    Taylor and power series
    Intervals and radii of convergence

LECTURES AND TEXTBOOK
    Algebra and calculus of power series

WORKBOOK
    Worksheet 6F.4. Operations with power series[‡]

OTHER PART 6 WORKBOOK ASSIGNMENTS

THE ϵN- AND YN-DEFINITIONS OF CONVERGENCE AND DIVERGENCE TO $\pm\infty$[†]
    Worksheet 6C.2. The ϵN- and YN-definitions of convergence and divergence to ∞[†]

TAYLOR'S THEOREM
    Worksheet 6F.3. Error estimates vs. actual errors for Taylor polynomials

---

[†]Written to be at least partially completed in supervised workshop sessions.
[‡]Suitable for homework or workshop sessions

Graphing calculator workbook   1

Worksheet 1A.1[†]

Maximizing an area: Initial investigations

Name _____ Date _____

Others in your group _____

Instructor, Teaching Assistant, and/or Recitation section _____

Directions   *Put first drafts of your calculations and answers on scratch paper. Take your time, work carefully, and discuss your solution with at least one other student before putting a final draft on this sheet or on other paper. Turn in all your work.*

Problem 1   You want to build a rectangular organic garden using a dormitory wall as one side and up to 34 feet of fence for the three other sides (Figure 1). Calculate the area of the garden and the length of fence used for various choices of the width w and length L. Experiment until you find the dimensions that give as large an area as you can determine. Put the results of several calculations which support your conclusion in the table below the sample. Do not use any calculus.

FIGURE 1

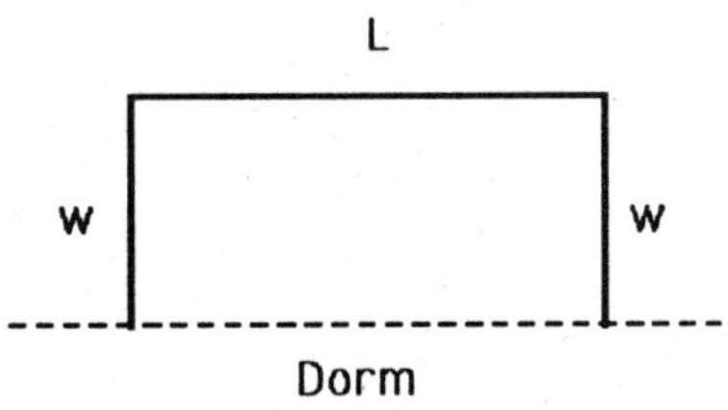

Conclusion:

Representative calculations

	w	L	FENCE LENGTH (≤ 34)	AREA
0.	9	15	33	135
1.				
2.				
3.				
4.				
5.				
6.				
7.				
8.				

[†] In this worksheet you will use arithmetic and trial and error to investigate a type of problem that you will be able to solve later with calculus. This worksheet and Worksheet 1E.1 were suggested to the author by Bill Helton at UCSD.

Graphing calculator workbook

Worksheet 1A.2[†]

Minimizing a length: Initial investigations

Name _____ Date _____

Others in your group _____

Instructor, Teaching Assistant, and/or Recitation section _____

Directions Put first drafts of your calculations and answers on scratch paper. Take your time, work carefully, and discuss your solution with at least one other student before putting a final draft on this sheet or on other paper. Turn in all your work.

Problem 1 You and your housemates want to build a four-sided pen for your dog using the 20-foot wide end of the house as one side. Some of you want a trapezoid-shaped pen as in Figure 1. Others want a rectangular pen as in Figure 2. You finally agree to build the pen that uses the least amount of fence, provided that it has an area of at least 160 square feet. Calculate the area of the pen and the total length of fence needed for various choices of the lengths x and y of the sides perpendicular to the house. Experiment until you find the dimensions that use as small a length of fence as you can determine. Put your conclusion on the back of this sheet and give several calculations which support it in the table below. Do not use calculus.

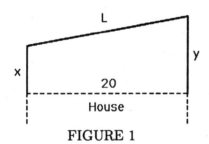
FIGURE 1

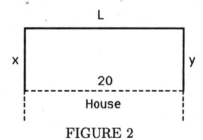
FIGURE 2

Representative calculations[‡]

	x	y	L	AREA (≥ 160)	FENCE LENGTH
0.	7	11	$\sqrt{20^2 + 4^2} = \sqrt{416}$	180	$18 + \sqrt{416} \doteq 38.4$
1.					
2.					
3.					
4.					
5.					
6.					

[†] In this worksheet you will use arithmetic and trial and error to investigate a type of problem that you will be able to solve later with calculus.

[‡] Equal signs are used in this workbook for exact equations; the symbol $\doteq$ is used in equations involving finite decimals that are correct except for round-off errors; and $\approx$ is used for other approximations.

4   Graphing calculator workbook

Conclusion:

Problem 2   If you think you know what choices of x and y require the least amount of fence,
prove that your conclusion is correct—using geometry, not calculus.

Graphing calculator workbook

Worksheet 1A.3[†]

Minimizing and maximizing an area: Initial investigations

Name _____ Date _____

Others in your group _____

Instructor, Teaching Assistant, and/or Recitation section _____

Directions Put your initial calculations and first drafts of your conclusions on scratch paper. Take your time, work carefully, and give your conclusions in well written sentences. Discuss your solution with at least one other student before putting a final draft on this sheet or other sheets of paper, as needed. Turn in all your work.

Problem 1 Your boss at Circle in the Square Pizza Parlor has decided to celebrate his sixtieth-fourth birthday with a promotion during which he will sell one square and one round piece of pizza for $9.98, with the guarantee that the total perimeter of the two pieces will be at least 64 inches (Figure 1). He asks you what size pieces to make to minimize the area of pizza and maximize his profits. Calculate the total area and perimeter of the square and circle for various choices of width w of the square and radius r of the circle. Experiment until you find the dimensions that give the smallest total area that you can determine with the requirement that the total perimeter be ≥ 64. Put your conclusion on the back of this sheet and list the results of several calculations that support it in the table below. Do not use calculus.

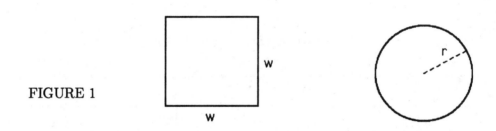

FIGURE 1

Representative calculations

	w	r	TOTAL PERIMETER (≥ 64)	TOTAL AREA
0.	10	4	$40 + 8\pi \doteq 65.1$	$100 + 16\pi \doteq 150.3$
1.				
2.				
3.				
4.				
5.				

[†] In this worksheet you will use arithmetic and trial and error to investigate a type of problem that you will be able to solve later with calculus.

Conclusion:

Problem 2 Find dimensions that give the largest total area of pizza you can determine under the requirement that the total perimeter be ≤ 64 inches.

Conclusion:

Representative calculations

	w	r	Total perimeter (≤ 64)	Total area
1.				
2.				
3.				
4.				
5.				
6.				

Graphing calculator workbook

Worksheet 1A.4[†]

Finding the closest point on a curve: Initial investigations

Name _____ Date _____

Others in your group _____

Instructor, Teaching Assistant, and/or Recitation section _____

Directions Put first drafts of your calculations and answers on scratch paper. Take your time, work carefully, and discuss your solution with at least one other student before putting a final draft on this sheet or on other paper. Turn in all your work.

Problem 1 Figure 1 shows the point $(1,1)$ and a point (x,y) on the curve $y = \dfrac{16}{x^2}$. Calculate the distance between these two points for various choices of x. Use eight decimal place accuracy. Experiment until you locate a point on $y = \dfrac{16}{x^2}$ that is as close as you can find to $(1,1)$. Put your conclusion on the back of this sheet and list the results of several calculations that support it in the table below, under the sample. Do not use calculus.

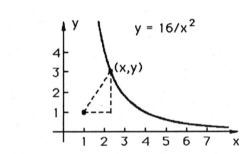

FIGURE 1

Representative calculations

	x	$y = 16/x^2$	DISTANCE
0.	2.3	$\doteq 3.024574669$	$\sqrt{(2.3-1)^2 + (3.024574669-1)^2} \doteq 2.406013839$
1.			
2.			
3.			
4.			
5.			
6.			

[†] In this worksheet you will use arithmetic and trial and error to investigate a type of problem that you will be able to solve later with calculus.

Conclusion:

Graphing calculator workbook   9

Calculator instructions 1B.1: Basic operations

Texas Instruments TI-81[†]

If you buy your calculator new, open its battery cover to be sure it has a strapped-in, pill-shaped battery, which enables you to change the four AAA flashlight batteries without erasing stored programs. Otherwise, you might be able to change the four AAA batteries without erasing programs if you turn the calculator off, put on its lid, and replace the batteries quickly, one at a time. Change the batteries as soon as the screen dims when graphs are generated.

To turn the calculator on, press the ON key. Press 2nd followed by the up cursor key ▲ to increase display contrast and by ▼ to decrease it. Press the MODE key. The screen should show

Norm	Sci Eng
Float	0123456789
Rad	Deg
Function	Param

Connected	Dot
Sequence	Simul
Grid Off	Grid On
Rect	Polar

with the boxed words highlighted (printed in white on a black background). If the wrong item is highlighted on any line, use the cursor keys to move the flashing box to the correct item and press ENTER . **Norm** denotes normal notation, **Sci** is scientific notation, and **Eng** is engineering notation for decimals. **Float** causes decimals to be printed with 10 digits. Choosing an integer instead of **Float** causes that many digits be shown after decimal points. **Rad** is for radians and **Deg** is for degrees. The other selections will be explained later when we discuss graphing.

The CLEAR key either clears an entry or returns you to a previous screen. 2nd activates the blue commands above the keys. 2nd QUIT returns you to the home screen where calculations are made. If you make an error in a command or calculation, the type of error is given and you have the choice of pressing 1 to return to where the error was commited to fix it, or 2 to cancel the incorrect command.

Read the parts of the owner's manual dealing with operating the calculator and on generating graphs of functions. Then work the following examples.

Example 1   Calculate the approximate decimal value of $\sqrt{6}\,\cos(9/7)$.

Solution   Press 2nd $\sqrt{\ }$ 6 cos (9 ÷ 7) to have the screen read $\sqrt{\ }\,6\cos\,(9/7)$. Then press ENTER to get the answer 0.6888851432. You do not need a multiplication sign after the 6 (as you would on a computer) because the calculator interprets juxtaposition of the digit 6 and the symbol cos as multiplication. If your answer is 2.448873047, your calculator is in degree mode.

[†]Substitutes of all calculator-specific pages are available for Texas Instruments TI-82 and TI-85, Casio fx-7000G and fx-7700G, and Sharp EL-9200 and EL-9300 graphing calculators and for the Macintosh software Calculus T/L.

10   Graphing calculator workbook

Priority of operations

The calculator generally follows standard algebraic notation when dealing with parentheses, negation, sums, differences, products, quotients, and exponents. Expressions inside parentheses are evaluated first and, in each expression, function evaluations are generally performed first. Addition, subtraction, and negation are not performed until another addition, subtraction, or the end of the expression is reached. Multiplication and division are not performed until another multiplication, division, addition, subtraction, or the end of the expression is reached.[†] Exponentiation is carried out as soon as another exponentiation, multiplication, division, addition, subtraction, or the end of the expression is reached.

For example, to evaluate $8+2\times3^{\sqrt{4}}-1$, enter $8+2*3 \wedge \sqrt{\ } 4-1$, with $\wedge$ for exponentiation, and press $\boxed{\text{ENTER}}$. The calculator does the arithmetic in five steps—without displaying intermediate results. First the square root of 4 is found, giving $8 + 2*3 \wedge 2 - 1$. Next, the power of 3 is calculated, yielding $8 + 2 * 9 - 1$. The multiplication is carried out, giving $8 + 18 - 1$. The addition is performed, giving $26 - 1$. And, finally, the subtraction is carried out, producing the final answer 25.

Example 2   Edit the expression in Example 1 to calculate $6\cos(\pi/7^3)$.

Solution   If you cleared $\sqrt{\ }6\cos(9/7)$ from the screen, put it back. Press $\boxed{\blacktriangle}$ and/or $\boxed{\blacktriangleleft}$ until the cursor is under the square root sign and press $\boxed{\text{DEL}}$ to delete it. Move the cursor to the 9 and press $\boxed{\text{2nd}}$ $\boxed{\pi}$ to replace the 9 with π. Put the cursor on the close parenthesis and press $\boxed{\text{INS}}$. Press $\boxed{\wedge}$ $\boxed{3}$ to insert $\wedge$ 3 before the close parenthesis, so the screen reads $6\cos(\pi/7\wedge3)$, and press $\boxed{\text{ENTER}}$ for the answer 5.999748331.

Example 3   Calculate $(-5 - 1.63 \times 10^{-2})^{-1}$.

Solution   The key $\boxed{\text{EE}}$ followed by a positive or negative integer n gives $\times 10^n$, which is displayed as E n. Press $\boxed{(}$ $\boxed{(-)}$ $\boxed{5}$ $\boxed{-}$ $\boxed{1}$ $\boxed{.}$ $\boxed{6}$ $\boxed{3}$ $\boxed{\text{EE}}$ $\boxed{(-)}$ $\boxed{2}$ $\boxed{)}$ $\boxed{x^{-1}}$ to display $(-5-1.63\text{E-}2)^{-1}$. Press $\boxed{\text{ENTER}}$ for the answer -0.1993501186. Notice that $\boxed{(-)}$ is for negation, $\boxed{-}$ is for subtraction, and $\boxed{x^{-1}}$ is for taking reciprocals.

Storing values

The key $\boxed{\text{ALPHA}}$ puts the calculator in alpha mode, activating the grey letters and other symbols above the keys. Pressing $\boxed{\text{2nd}}$ $\boxed{\text{A-LOCK}}$ locks the calculator in alpha mode. Pressing $\boxed{\text{ALPHA}}$ or $\boxed{\text{ENTER}}$ takes you out of alpha mode. The letters are names of memory cells. Entering a number followed by $\boxed{\text{STO▶}}$, the letter, and $\boxed{\text{ENTER}}$ assigns

---

[†]There is one important exception to this rule: multiplication that is indicated by juxtaposing symbols without a $*$ is performed before multiplication denoted with $*$ and before division. For example, $5 \div 3\sin(0.1)$ is interpreted as $5/[3\sin(0.1)] \doteq 16.69$, while $5 \div 3 * \sin(0.1)$ is interpreted as $(5/3)\sin(0.1) \doteq 0.166$.

Graphing calculator workbook

that number to the letter. The number can then be recalled with ALPHA followed by the letter.

Example 4 Calculate $A + BC^D$ for $A = 7$, $B = 6$, $C = 5$, and $D = 4$.

Solution Press 7 STO▶ A ENTER to assign A the value 7. Notice that the ALPHA key is not needed here. The calculator is automatically in alpha mode immediately after STO▶ is pressed.

Press 6 STO▶ B ENTER ; 5 STO▶ C ENTER ; 4 STO▶ D ENTER to assign the values to the other letters. Then ALPHA A + ALPHA B ALPHA C ∧ ALPHA D puts $A + BC \wedge D$ on the screen. ENTER gives the answer 3757.

Adding parentheses

Often extra parentheses have to be used to express numerators, denominators, and exponents on the calculator.

Example 5 Evaluate $\dfrac{2+8}{4-6} - 3^{5-1}$ without simplifying the expression first.

Solution Use (2 + 8) ÷ (4 - 6) - 3 ∧ (5 - 1) to display $(2+8)/(4-6) - 3 \wedge (5-1)$. Press ENTER for the answer -86.

Parentheses should also be used to show the order of operations whenever functions are combined, since the calculator's rules for evaluating combinations of functions without parentheses may not be what you expect. For example, sin (2) x^2 puts $\sin(2)^2$ on the screen and yields $\sin(2^2) = \sin(4) \doteq -0.7568024953$ rather than $(\sin 2)^2 \doteq 0.8268218104$ as you might expect. To evaluate $(\sin 2)^2$, use (sin 2) x^2.

More examples

Work the following examples to be sure you understand the calculator's rules. Do not look at the solutions except to find your errors.

Example 6 $4 - 3\sin(\pi/4) \doteq 1.878679656$

Solution 4 − 3 sin (2nd π ÷ 4) ENTER

Example 7 $\dfrac{1.34 \times 10^6 - 4 \times 10^5}{7.12 \times 10^{-8}} \doteq 1.320224719 \times 10^{13}$

Solution (1 . 3 4 EE 6 − 4 EE 5) ÷ (7 . 1 2 EE (-) 8) ENTER

Example 8 $\sqrt{4 + 7^{8-10}} \doteq 2.00509555$

Solution 2nd √ (4 + 7 ^ (8 − 1 0)) ENTER

Example 9 For $A = 1.234$: $100 \sin^2(A^{1.5}) \doteq 96.05302604$. (Remember that $\sin^2 x$ denotes $(\sin x)^2$.)

Solution 1 . 2 3 4 STO▶ A ENTER
1 0 0 (sin (ALPHA A ^ 1 . 5)) x^2 ENTER

Example 10 $(-32)^{4/5} = 16$ (This one is tricky because of idiosyncrasies of the calculator. It will evaluate $x^{1/n}$ for negative x and odd integers n in apparently every situation, but will not always accept $x^{m/n}$ with negative x, odd n, and m an integer > 1. It is generally best to write $x^{m/n}$ as $(x^m)^{1/n}$ for odd n.)

Solution (-) 3 2 ^ (4 ÷ 5) ENTER

gives the wrong result since the negation is perfomed after the power.

((-) 3 2) ^ (4 ÷ 5) ENTER

results in an error message since the calculator cannot evaluate the four-fifths power of the negative number.

((-) 3 2) ^ 4 ^ (1 ÷ 5) ENTER

or

((-) 3 2) ^ (1 ÷ 5) ^ 4 ENTER

will work, since the calculator will find the one-fifth power of any real number.

Graphing calculator workbook

Calculator instructions 1B.2: Generating graphs of functions

Texas Instruments TI-81

Press the ☐MODE☐ key. The screen should read

Norm Sci Eng	**Connected** Dot
Float 0123456789	**Sequence** Simul
Rad Deg	**Grid Off** Grid On
Function Param	**Rect** Polar

with the boxed words highlighted (printed in white on a black background). If the wrong item is highlighted on any line, move the flashing box to the correct item and press ☐ENTER☐. **Function** is for generating graphs $y = f(x)$ of functions. **Param** (parametric) is for curves with x and y given as functions of a third variable t.

A graph $y = f(x)$ is generated from points (x, y), where x is the horizontal coordinate of a pixel (dot) on the screen and y is the vertical pixel coordinate that is closest to the corresponding value $f(x)$ of the function. The points are joined by lines if **Connected** is chosen and are not joined if **Dot** is chosen. Use **Sequence** to have two, three or four graphs drawn one after the other, and **Simul** (simultaneous) to have them drawn at the same time. If **Grid On** were chosen, dots would be shown at the values of x and y corresponding to the ticmarks on the axes. **Rect** is for rectangular and **Polar** for polar coordinates. Press ☐ON☐ to interrupt the drawing of a graph.

Example 1   Generate the curves $y = x^2 - x$ and $y = \dfrac{1}{x^2 + 1}$ for $-1 \le x \le 2$, $-0.5 \le y \le 1.5$ on the same screen.

Solution   Press ☐RANGE☐ and enter the values x-min $= -1$, x-max $= 2$, x-scl $= 1$, y-min $= -0.5$, y-max $= 1.5$ and y-scl $= 1$. x-res should have the value 1. Setting x-scl and y-scl equal to 1 causes the ticmarks to be one unit apart on both axes. Use ☐▲☐ ☐▼☐ ☐◄☐ and ☐►☐ to move the cursor and ☐DEL☐ to delete extraneous symbols.

Next press ☐Y =☐. The cursor should be after : $Y_1 = $. Press ☐CLEAR☐ to remove any previous formula for Y_1 and then ☐X|T☐ ☐x^2☐ ☐−☐ ☐X|T☐ to have the first line read : $Y_1 = X^2 - X$. (☐X|T☐ is the same as ☐ALPHA☐ ☐X☐ in **Function** mode and the same as ☐ALPHA☐ ☐T☐ in **Param** mode.) If the equal sign is not highlighted, move the cursor to it and press ☐ENTER☐ to highlight it. Move the cursor to after the equal sign on the second line, press ☐CLEAR☐ and then ☐1☐ ☐÷☐ ☐(☐ ☐X|T☐ ☐x^2☐ ☐+☐ ☐1☐ ☐)☐, so the second line reads : $Y_2 = 1/(X^2 + 1)$. Be sure the first and second equal signs are highlighted and the third and fourth equal signs are not.

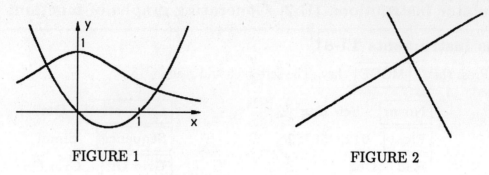

FIGURE 1   FIGURE 2

Finally, press $\boxed{\text{GRAPH}}$. Because the first two equal signs are highlighted, the parabola $y = x^2 - x$ will be drawn, followed by "Agnesi's versiera" $y = 1/(x^2+1)$, as in in Figure 1. Press $\boxed{\text{ON}}$ to interrupt the generation of graphs.

Example 2   Use the trace and zoom commands twice to find the approximate coordinates of each of the two intersections of the curves in Example 1.

Solution   If the curves are not still on screen, generate them again. Press $\boxed{\text{TRACE}}$. A flashing × will appear on one of the curves. Press $\boxed{\blacktriangleleft}$ or $\boxed{\blacktriangleright}$ until the × is more or less at the intersection of the curves to the left of the y-axis, and the screen shows a value of X close to -0.5. (Pressing $\boxed{\blacktriangle}$ or $\boxed{\blacktriangledown}$ causes the × to move from one curve to the other.) Press $\boxed{\text{ZOOM}}$ $\boxed{4}$ and set the zooming factors to XFact = 10 and YFact = 10. Press $\boxed{\text{ENTER}}$ to complete each entry. Press $\boxed{\text{ZOOM}}$ $\boxed{2}$ $\boxed{\text{ENTER}}$ and the two curves will be redrawn in a rectangle with its center at the position of the × and with x- and y-ranges that are one-tenth of those in the previous graph, more or less as in Figure 2. Press $\boxed{\text{TRACE}}$ and move the flashing × to the intersection. Press $\boxed{\text{ZOOM}}$ $\boxed{2}$ $\boxed{\text{ENTER}}$ to redraw the curves in another, smaller rectangle. Press $\boxed{\text{TRACE}}$, move the × to the intersection, and read the coordinates of the × from the bottom of the screen. You should get values close to $X = -0.519, Y = 0.788$.

Set the range back to $-1 \leq x \leq 2$, $-0.5 \leq y \leq 1.5$ and press $\boxed{\text{GRAPH}}$ to regenerate the curves as in Figure 1. Repeat the above process to find the approximate coordinates $X = 1.291, Y = 0.375$ of the intersection to the right of the y-axis.

The box command described in the owner's manual gives another—graphical and less precise—way of zooming in on graphs.

Evaluating functions in the Y = menu

To find the approximate decimal value at x of the function stored as Y_1 in the $Y =$ menu, store x in memory X by putting an expression for it in the home screen and pressing $\boxed{\text{STO}\blacktriangleright}$ $\boxed{X|T}$ $\boxed{\text{ENTER}}$. Then press $\boxed{\text{2nd}}$ $\boxed{\text{Y-vars}}$ $\boxed{1}$ $\boxed{\text{ENTER}}$. Values of Y_2, Y_3, and Y_4 can be found similarly.

Graphing calculator workbook

Calculator instructions 1B.3: Tips for solving worksheets

Texas Instruments TI-81

This section contains suggestions for solving worksheets that require or can be more easily solved with special calculator techniques.

Worksheet 1C.1

Use $-9 \leq x \leq 9$, $-5 \leq y \leq 7$ or other x- and y-ranges for which (y-max $- y$-min) $= \frac{2}{3}(x$-max $- x$-min), so there are equal scales on the axes and, consequently, right angles appear as right angles.

Worksheet 1D.2

(1) Set $Y_1 = (1/4)X \wedge 4 + X - 10$ or $Y_1 = X \wedge 4/4 + X - 10$ in the Y= menu, and be sure only the first equal sign is highlighted. Do not use $Y_1 = 1/4 X \wedge 4 + X - 10$. This puts $X \wedge 4$ in the denominator, because the multiplication by juxtapositioning of symbols without a times sign is calculated before the division. (2) For columns 3 and 4, put an exact expression for each x in the home screen and press $\boxed{\text{STO}\blacktriangleright}$ $\boxed{X|T}$ $\boxed{\text{ENTER}}$ to store it in memory X and display its decimal value to ten digits. Then find the decimal value of y with $\boxed{\text{2nd}}$ $\boxed{\text{Y-VARS}}$ $\boxed{1}$ $\boxed{\text{ENTER}}$. Use the trace command to check your values with the graph.

Worksheet 1D.3

Set $Y_1 = X^2 + 1/(1 - X)$ in the Y= menu, and be sure only the first equal sign is highlighted. See the tips for Problem 2 of Worksheet 1D.2.

Worksheet 1D.4

Set $Y_1 = \sqrt{\ } X/(X^2 + 1)$ in the Y= menu, and be sure only the first equal sign is highlighted. See the tips for Problem 2 on Worksheet 1D.2.

Worksheets 1E.1–1E.4

After you generate the graph, press $\boxed{\text{TRACE}}$. Watch for the least or greatest traced y-value as you use $\boxed{\blacktriangleright}$ and $\boxed{\blacktriangleleft}$ to move the × along the graph. In Problem 2f of Worksheet 1E.4, enter y-min with $\boxed{(-)}$ $\boxed{0}$ $\boxed{.}$ $\boxed{5}$ $\boxed{\text{EE}}$ $\boxed{(-)}$ $\boxed{8}$ $\boxed{\text{ENTER}}$, etc.

Worksheet 1G.1

(1a–c) Set $Y_1 = (X \wedge 4 - 1)/(X^2 - 1)$ in the Y= menu. There are tiny gaps in the graphs at $x = 1$ where the function is not defined because with these choices of x-ranges there are pixels at $x = 1$. (1d) Follow the tips for Problem 2 of Worksheet 1D.2. (2a) Set $Y_2 = 1.5, Y_3 = 2$, and $Y_4 = 2.5$. (2d) Set $Y_2 = 1.999, Y_3 = 2$, and $Y_4 = 2.001$.

Worksheet 1G.2

(1) Set $Y_1 = (X - 2)/\text{abs}(X - 2)$. The two lines that form the graph are not joined at $x = 2$ because with $-1 \leq x \leq 4$ there is a pixel exactly at $x = 2$. With other x-ranges, such as $-2 \leq x \leq 4$, there is no pixel exactly at $x = 2$ and an extraneous line is drawn joining the two parts. This line can be avoided with any x-range by selecting **DOT** in the Mode menu.

(2) Use $Y_1 = \sqrt{\ }(X^2 - 1)/(X - 1)$.

Worksheet 1G.3

(1) For the table either use $Y_1 = -3 - 3/X$ for $x < 0$ and $Y_2 = 3 - 3/X$ for $x > 0$ or use $Y_1 = (-3 - 3/X)(X < 0) + (3 - 3/X)(X > 0)$. Press $\boxed{\text{2nd}}$ $\boxed{\text{TEST}}$ $\boxed{5}$ for $<$ and $\boxed{\text{2nd}}$ $\boxed{\text{TEST}}$ $\boxed{3}$ for $>$. (2) Setting $Y_1 = X \wedge (2/3)$ will give the portion of $y = x^{2/3}$ for $x \geq 0$, which is all you need in this problem. Setting $Y_1 = (X^2) \wedge (1/3)$ will also give the portion for $x < 0$. Use $Y_2 = 10$ for part (a), $Y_2 = 500$ for part (b), and $Y_2 = 1000$ for part (c).

Worksheet 1H.1

(2) Extraneous vertical lines are drawn at $x = 0.25$ and 0.75 because no pixels have those exact x-coordinates. The extraneous lines do not appear with $0 \leq x \leq 1.25$ because then there are pixels at $x = 0.25, 0.5, 0.75$, and 1 (20%, 40%, 60% and 80% across the screen).

Worksheets 1I.2 and 1I.3

Generate the graphs with the given x-range and with a y-range that you think might be good. If the choice of y-range is poor so that no curve is visible or you cannot tell from the curve that is generated how to improve the plot, see what values of y are involved—even if no curve is shown—by using $\boxed{\text{TRACE}}$ followed by $\boxed{\blacktriangleleft}$ and $\boxed{\blacktriangleright}$ to have the displayed values of x run from x-min to x-max. Use the displayed values of y to select a suitable y-range.

Worksheet 2C.2

Press $\boxed{\text{MODE}}$, put the cursor on **2**, and press $\boxed{\text{ENTER}}$ to have values shown with two digits after the decimal point. Restore ten-digit displays by setting the mode screen back to **Float** after you finish this worksheet.

Worksheet 2J.1

Enter $F(r)$ and $\dfrac{dF}{dr}(r)$ in the $Y=$ menu as Y_1 and Y_2, respectively, with X in place of r. Run the program and press $\boxed{\text{ENTER}}$ for each step. Press $\boxed{\text{ON}}$ $\boxed{2}$ to abort the program when the values of x_j repeat.

Worksheets 2J.2–2J.4

See the tips for Worksheet 2J.1.

Worksheet 3C.1

Press $\boxed{\text{MODE}}$, put the cursor on **6**, and press $\boxed{\text{ENTER}}$ to have decimal values shown with six digits after the decimal point. Restore ten-digit displays by setting the mode screen back to **Float** after you finish this worksheet.

Worksheet 3C.2

Follow the tips for Worksheet 3C.1 with 2 instead of 6.

Worksheet 3D.1

Follow the tips for Worksheet 3C.1 with 4 instead of 6.

Graphing calculator workbook

Worksheet 4A.1

(1b–d) Put $(2 \wedge 0.0001 - 1)/0.0001$ in the home screen and press $\boxed{\text{ENTER}}$ for the value of (1) at $b = 2$. Then press $\boxed{\blacktriangle}$ to recall the expression and edit it, with $\boxed{\text{INS}}$ and $\boxed{\text{DEL}}$ as needed, to replace 2 by the other values of b. (2) Set $Y_1 = e \wedge X$ and $Y_2 = (e \wedge (X + 0.0001) - e \wedge X)/0.0001$. Put each value of x in the home screen, press $\boxed{\text{STO}\blacktriangleright}$ $\boxed{X|T}$ $\boxed{\text{ENTER}}$ followed by $\boxed{\text{2nd}}$ $\boxed{\text{Y-VARS}}$ $\boxed{1}$ $\boxed{\text{ENTER}}$ and $\boxed{\text{2nd}}$ $\boxed{\text{Y-VARS}}$ $\boxed{2}$ $\boxed{\text{ENTER}}$ for the values of the two functions.
(3) Use $Y_1 = 10 \wedge X$ and $Y_2 = (10 \wedge (X + 0.001) - 10 \wedge X)/0.001$
(4) Use $Y_1 = 2 \wedge X$ and $Y_2 = (2 \wedge (X + 0.001) - 2 \wedge X)/0.001$

Worksheet 4A.2

(2a) Put $(\ln(1 + 0.000001) - \ln 1)/0.000001$ in the home screen and press $\boxed{\text{ENTER}}$ for the value of the difference quotient at $x = 1$. Then press $\boxed{\blacktriangle}$ to recall the expression and edit it to calculate the difference quotient for the other values of x. (3a) Follow the tips for Problem 2 with log instead of ln.

Worksheet 4B.3

Enter function (2) as $Y_1 = P(1 + R/N) \wedge \text{Int} (NX)$. Press $\boxed{\text{MATH}}$ $\boxed{\blacktriangleright}$ $\boxed{4}$ for Int. Enter function (3) as $Y_2 = Pe \wedge (RX)$. Press $\boxed{\text{MODE}}$, put the cursor on **Dot**, and press $\boxed{\text{ENTER}}$ to put the calculator in dot mode. Put the value of P in the home screen and press $\boxed{\text{STO}\blacktriangleright}$ $\boxed{P}$ $\boxed{\text{ENTER}}$ to store it as P. Store the values of R and N similarly. Then set the x- and y-ranges and generate the graphs you need. Set the Mode menu back to **Connected** when you have finished this worksheet.

Worksheet 6C.1

Set $Y_1 = 6/(1+J)$ for Problem 1 and the other given formulas for Problems 2–9. Run Sequence program 6B with $J0 = 1$ and MAX $J = 24$. Give $K = 0$ at the end of the program to have no other graphs generated and finish the program, or press $\boxed{\text{ON}}$ $\boxed{2}$ to abort it at any time.

Worksheet 6C.2

(1) Use $Y_1 = 2 + \cos(J\pi/3)/J^2$, $Y_2 = 2.1$, and $Y_3 = 1.9$ in part (a). Use $Y_2 = 2.01, Y_3 = 1.99$ in part (b) and use $Y_2 = 2.001, Y_3 = 1.999$ in part (c). (2)–(3) Adapt the tips for Problem 1.

Worksheet 6E.1

(1a) Set $Y_1 = 0.6 \wedge J/(J+1)$ or $Y_1 = (1/(J+1))(0.6 \wedge J)$ and use $\boxed{\text{2nd}}$ $\boxed{\text{DRAW}}$ $\boxed{1}$ $\boxed{\text{ENTER}}$ to clear the graphics screen before running the program. Use $L = 1, J0 = 0$, and Max$J = 12$ in the program and ask to see the partial sums after $N = -1$, so they are all displayed. Set $K = 0$ at the end of the program so no other curves are generated. Use $\boxed{\text{ON}}$ $\boxed{2}$ to abort the program at any other time. (1b) Set $Y_2 = 0.6 \wedge J$, clear the graphics screen, and run the program twice, first with $L = 1$ and then with $L = 2$. (2)–(4) Adapt the tips for Problem 1.

Worksheet 6E.2

(1a) Set $Y_1 = (-0.9) \wedge J$ and use $\boxed{\text{2nd}}$ $\boxed{\text{DRAW}}$ $\boxed{1}$ $\boxed{\text{ENTER}}$ to clear the graphics screen before running the program. Use $L = 1, J0 = 0$, and $\text{Max}J = 12$ in the program and ask to see the partial sums after $N = -1$, so they are all displayed. Set $K = 0$ at the end of the program so no other curves are generated. Use $\boxed{\text{ON}}$ $\boxed{2}$ to abort the program at any other time. **(1b)** Use $L = 1$, $J0 = 0$, and Max $J = 62$ in Program 6B. Ask to see the partial sums after $N = 62$ so none are displayed. **(2)–(4)** Adapt the tips for Problem 1.

Worksheet 6E.3

(1) Set $Y_1 = \cos(J^2)/(J \wedge +1)$ and use $\boxed{\text{2nd}}$ $\boxed{\text{DRAW}}$ $\boxed{1}$ $\boxed{\text{ENTER}}$ to clear the graphics screen before running the program. Use $L = 1, J0 = 0$, and $\text{Max}J = 20$ in the program, and ask to see the partial sums after $N = 16$. Set $K = 0$ at the end of the program so no other curves are generated. Use $\boxed{\text{ON}}$ $\boxed{2}$ to abort the program at any other time. **(2)–(7)** Adapt the tips for Problem 1.

Worksheet 6F.2

(1) Use $Y_1 = \cos X$, $Y_2 = P_n(X)$, and $Y_3 = Y_1 - Y_2$ with the appropriate formula for each n. The keys $\boxed{\text{MATH}}$ $\boxed{5}$ give ! (factorial). Highlight only the first two equal signs to generate the required graphs. Store values as X by putting them in the home screen and pressing $\boxed{\text{STO}\blacktriangleright}$ $\boxed{\text{ENTER}}$. Press $\boxed{\text{2nd}}$ $\boxed{\text{Y-VARS}}$ followed by $\boxed{1}$ $\boxed{2}$ or $\boxed{3}$ and $\boxed{\text{ENTER}}$ for the numbers in columns 2 and 3 of Table 1.

Worksheets 6F.3 and 6F.4

Adapt the tips for Worksheet 6F.2.

Graphing calculator workbook

Worksheet 1C.1[†]

Can you match this?

Name _____ Date _____

Others in your group _____

Instructor, Teaching Assistant, and/or Recitation section _____

Directions Put first drafts of your calculations and answers on scratch paper. Take your time, work carefully, and discuss your solution with at least one other student before putting a final draft on this sheet or on other paper. Turn in all your work.

Problem 1 Figure 1 below shows a design formed by four lines in an xy-plane with equal scales on the x- and y-axes. The line through the origin and the point P has slope less than 1, there is a right angle at P, and the design is symmetric about the y-axis. Find equations of four lines which satisfy these conditions. (There are an infinite number of possible solutions). Check your results by generating the four lines on your calculator or computer.

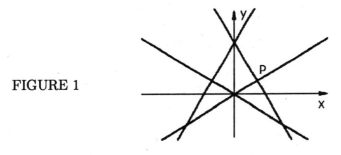

FIGURE 1

Equations:

[†]This worksheet is a review of equations of lines. See Calculator instructions 1B.3 for tips on using a calculator in solving it.

Graphing calculator workbook

Worksheet 1C.2[†]

Pick a friend

Name _____ Date _____

Others in your group _____

Instructor, Teaching Assistant, and/or Recitation section _____

Directions Put first drafts of your calculations and answers on scratch paper. Take your time, work carefully, and discuss your solution with at least one other student before putting a final draft on this sheet or on other paper. Turn in all your work.

Problem 1 The face in Figure 1 was drawn in an xy-plane with its origin at the center of the largest circle. The five circles and two semicircles have the following equations. Match the equations to the curves.

(a) $x^2 + y^2 = 400$

(b) $(x - 7)^2 + (y - 9)^2 = 9$

(c) $(x + 7)^2 + (y - 9)^2 = 9$

(d) $(x - 7 - 0.75\sqrt{2})^2 + (y - 9 + 0.75\sqrt{2})^2 = 2.25$

(e) $(x + 7 - 0.75\sqrt{2})^2 + (y - 9 + 0.75\sqrt{2})^2 = 2.25$

(f) $y = -6 - \sqrt{64 - x^2}$

(g) $y = \sqrt{4 - x^2}$

FIGURE 1

FIGURE 2

[†]This worksheet is a review of equations of circles and of semicircles with horizontal and vertical bases.

Problem 2   The circle in Figure 2 has the equation

$$x^2 + y^2 = 400. \tag{1}$$

The frown and nose are the semicircles

$$y = -14 + \sqrt{64 - x^2} \quad \text{and} \quad y = -\sqrt{9 - x^2}. \tag{3}$$

The left eye is formed by the semicircles

$$y = 8 + \sqrt{16 - (x + 6)^2} \quad \text{and} \quad y = 12 - \sqrt{16 - (x + 6)^2}. \tag{4}$$

The right eye is formed by

$$y = 8 + \sqrt{16 - (x - 6)^2} \quad \text{and} \quad y = 12 - \sqrt{16 - (x - 6)^2}. \tag{5}$$

The ears are

$$x = 19.5 + \sqrt{25 - y^2} \quad \text{and} \quad x = -19.5 - \sqrt{25 - y^2}. \tag{6}$$

a.   Why does the left eye have the equation

$$\left[8 + \sqrt{16 - (x + 6)^2} - y\right]\left[12 - \sqrt{16 - (x + 6)^2} - y\right] = 0? \tag{7}$$

b.   Why does the right eye have the equation

$$\left[8 + \sqrt{16 - (x - 6)^2} - y\right]\left[12 - \sqrt{16 - (x - 6)^2}\right] = 0? \tag{8}$$

c.   Explain why the entire face is given by

$$
\begin{aligned}
&\left[400 - x^2 - y^2\right]\left[-14 + \sqrt{64 - x^2} - y\right]\left[8 + \sqrt{16 - (x + 6)^2} - y\right] \\
&\times \left[12 - \sqrt{16 - (x + 6)^2} - y\right]\left[8 + \sqrt{16 - (x - 6)^2} - y\right] \\
&\times \left[12 - \sqrt{16 - (x - 6)^2} - y\right]\left[19.5 + \sqrt{25 - y^2} - x\right] \\
&\times \left[19.5 + \sqrt{25 - y^2} + x\right]\left[\sqrt{9 - x^2} + y\right] = 0.
\end{aligned}
\tag{9}
$$

Graphing calculator workbook

Worksheet 1D.1[†]

Pixels, decimals, coordinates, and angles

Name _____ Date _____

Others in your group _____

Instructor, Teaching Assistant, and/or Recitation section _____

Directions   *Put first drafts of your calculations and answers on scratch paper. Take your time, work carefully, and discuss your solution with at least one other student before putting a final draft on this sheet or on other paper. Turn in all your work.*

Problem 1   Use the calculator to complete the table below of ten-digit values of $5x^{1/3} = 5\sqrt[3]{x}$ at $x = -30, 0, 4, 6, 8$, and 10. To do these calculations quickly, set $Y_1 = 5X \wedge$ $(1/3)$ in the Y= menu. Put each x in the home screen and press $\boxed{\text{STO▶}}$ $\boxed{X|T}$ $\boxed{\text{ENTER}}$ to store it in memory X. Press $\boxed{\text{2nd}}$ $\boxed{\text{Y-VARS}}$ $\boxed{1}$ $\boxed{\text{ENTER}}$ for the value of Y_1. You will use these numbers in Problem 3.

   The value $5(-27)^{1/3} = 5(-3) = -15$ is exact, but -15.53616253 is only a decimal approximation of $5(-30)^{1/3}$, which cannot be represented by a finite decimal. Its value to 20 decimal places, for example, is -15.53616252976929433439. Which other of the y-values in the table do you recognize as exact?

x	$y = 5x^{1/3} \doteq$	x	$y = 5x^{1/3} \doteq$
-27	-15	-30	-15.53616253
4		8	
6		10	

Problem 2   The calculator screen is formed of square pixels that are darkened to display printing and graphs. The screen is 96 pixels wide and 64 pixels high. To see this, set x-min $= -47$, x-max $= 48$, x-scale $= 10$, y-min $= -31$, y-max $= 32$, y-scale $= 10$ and x-resolution $= 1$ in the Range menu. Press $\boxed{Y=}$. If any equal signs are highlighted, put the cursor on them and press $\boxed{\text{ENTER}}$ to unselect them so no graphs will be drawn. Press $\boxed{\text{GRAPH}}$ to see the graphics screen, and move the cross hairs around the screen with the cursor keys $\boxed{\blacktriangle}$ $\boxed{\blacktriangledown}$ $\boxed{\blacktriangleleft}$ $\boxed{\blacktriangleleft}$. The coordinates of the pixels at the cross hairs are displayed at the bottom of the screen. The pixels' coordinates should be integers and the pixels should be one unit apart in both the x- and y-directions.

   [†]Written for a TI-81 graphing calculator. The difference between exact and approximate decimal values of functions is observed in Problem 1. Problems 2 and 3 compare the numbers that are displayed when the cursor is moved around the graphics screen with and without the trace command. Problem 4 shows how changing scales can affect angles between displayed lines.

Graphing calculator workbook

a. Find choices of x-min, x-max, y-min, and y-max so that the pixels' coordinates are mutliples of 100 and the pixels are 100 units apart in the x- and y-directions.

b. Find choices of x-min, x-max, y-min, and y-max so that the pixels' coordinates are mutliples of 0.01 and the pixels are 0.01 units apart in the x- and y-directions.

c. Find choices of x-min, x-max, y-min, and y-max so that the pixels' x-coordinates are mutliples of 100 and their y-coordinates are multiples of 0.01.

Problem 3 Set x-min = -47, x-max = 48, x-scale = 10, y-min = -31, y-max = 32, and y-scale = 10 again. Press $\boxed{Y=}$ and highlight the equal sign for $Y_1 = 5X^{1/3}$ or re-enter the formula. Press $\boxed{GRAPH}$ to draw its graph and copy it in Figure 1. Move the cross hairs around with the cursor keys to see that the coordinates of the pixels have not changed from Problem 2. Then press $\boxed{TRACE}$ and the left and right cursor keys to move the x along the graph. Now what values are displayed and how do they relate to the coordinates of the pixels?

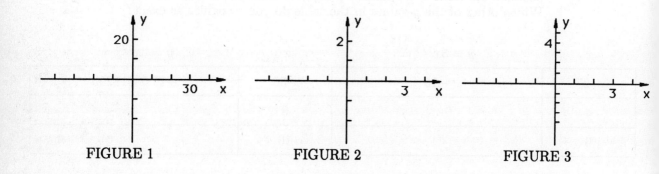

FIGURE 1   FIGURE 2   FIGURE 3

Problem 4 Set $Y_1 = X$ and $Y_2 = -X$ in the Y= menu. Use $\boxed{(-)}$ not $\boxed{-}$ for the negation sign and be sure only the first two equal signs are highlighted.

a. Set the ranges to $-4.7 \leq x \leq 4.8$, $-3.1 \leq y \leq 3.2$ with x-scale = 1, and y-scale = 1. Press $\boxed{GRAPH}$ to draw the two perpendicular lines $y = x$ and $y = -x$. Copy them in Figure 2. Use the cursor keys to move the cross hairs around the screen. How does this demonstrate that there are equal scales on the x- and y-axes?

b. Now change the y-range to $-6.2 \leq y \leq 6.4$, without changing the x-range, and generate the lines again. Copy them in Figure 3. Use the cursor keys to move the cross hairs around the screen. What does this tell you about the scales on the axes?

c. The lines in Figure 2 are perpendicular but the lines with the same equations in Figure 3 are not. Explain.

Graphing calculator workbook

Worksheet 1D.2[†]

Graphs and values of functions

Name _____ Date _____

Others in your group _____

Instructor, Teaching Assistant, and/or Recitation section _____

Directions   *Put first drafts of your calculations and answers on scratch paper. Take your time, work carefully, and discuss your solution with at least one other student before putting a final draft on this sheet or on other paper. Turn in all your work.*

Problem 1   Generate the graph of $y = \frac{1}{4}x^4 + x - 10$ on your calculator or computer with x-min $= -4$, x-max $= 4$, x-scale $= 1$, y-min $= -15$, y-max $= 20$, and y-scale $= 5$. Copy the graph in Figure 1.

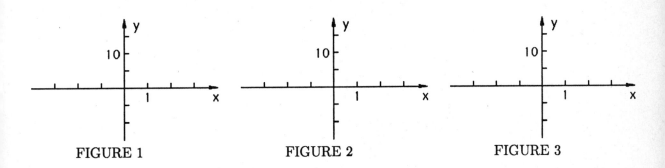

FIGURE 1   FIGURE 2   FIGURE 3

Problem 2   Find the exact values of $\frac{1}{4}x^4 + x - 10$ at $x = 0, \sqrt{2}, -3$, and π and complete the second column of the table below. Then put ten-digit decimal values of each x and y in the third and fourth columns. Check your values with the graph from Problem 1.

	Exact values	Ten-digit decimal values	
x	$y = \frac{1}{4}x^4 + x - 10$	$x \doteq$	$y \doteq$
0	$\frac{1}{4}(0)^4 + 0 - 10 = -10$	0	-10
$\sqrt{2}$	$\sqrt{2} - 9$	1.414213562	-7.585786438
-3			
π			17.49386541

[†]This worksheet gives practice finding values and generating graphs of functions and explores how changing a formula for a function changes its graph. See Calculator instuctions 1B.3 for tips on solving it with a calculator.

26   Graphing calculator workbook

Problem 3   Generate $y = \frac{1}{4}x^4 + x - 10$ and $y = \frac{1}{4}x^4 + x - 5$ together with the ranges from
Problem 1 and copy the two curves in Figure 2. How does changing 10 to 5 change
the curve and why?

Problem 4   Generate $y = \frac{1}{4}x^4 + x - 10$ and $y = \frac{1}{2}x^4 + x - 10$ together with the ranges from
Problem 1 and copy the two curves in Figure 3. How does changing $\frac{1}{4}$ to $\frac{1}{2}$ in the
equation change the curve and why?

Graphing calculator workbook

Worksheet 1D.3†

Graphs and values of functions

Name _____ Date _____

Others in your group _____

Instructor, Teaching Assistant, and/or Recitation section _____

Directions   Put first drafts of your calculations and answers on scratch paper. Take your time, work carefully, and discuss your solution with at least one other student before putting a final draft on this sheet or on other paper. Turn in all your work.

Problem 1   Generate the curve $y = x^2 + \dfrac{1}{1-x}$ on your calculator or computer with x-min $= -2$, x-max $= 3$, x-scale $= 1$, y-min $= -6$, y-max $= 8$, and y-scale $= 1$. Copy the graph in Figure 1.

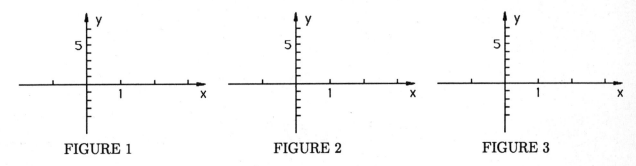

FIGURE 1   FIGURE 2   FIGURE 3

Problem 2   Find the exact values of $x^2 + \dfrac{1}{1-x}$ at $x = \tfrac{1}{3}$, 1.2, $\sqrt{3}$, and $\pi - 1$ and complete the second column of the table below. Then put ten-digit decimal values of each x and y in the third and fourth columns. Check your values by using the graph from Problem 1.

	EXACT VALUES		TEN-DIGIT DECIMAL VALUES
x	$y = x^2 + \dfrac{1}{1-x}$	$x \doteq$	$y \doteq$
$\tfrac{1}{3}$	$(\tfrac{1}{3})^2 + \dfrac{1}{1 - 1/3} = \tfrac{29}{18}$	0.333333333	1.611111111
1.2	-3.56		
$\sqrt{3}$	$3 + \dfrac{1}{1 - \sqrt{3}}$	1.732050808	1.633974596
$\pi - 1$			

---

†This worksheet gives practice finding values and generating graphs of functions and explores how changing a formula for a function changes its graph. See Calculator instuctions 1B.3 for tips on solving it with a calculator.

28   Graphing calculator workbook

Problem 3   Generate $y = x^2 + \dfrac{1}{1-x}$ and $y = x^2 + \dfrac{1}{2-x}$ together with the ranges from Problem 1 and copy the two curves in Figure 2. How does changing $\dfrac{1}{1-x}$ to $\dfrac{1}{2-x}$ change the curve and why?

Problem 4   Generate $y = x^2 + \dfrac{1}{1-x}$ and $y = x^2 - \dfrac{1}{1-x}$ together with the ranges from Problem 1 and copy the two curves in Figure 3. How does changing the plus sign to a minus sign change the curve and why?

Graphing calculator workbook

Worksheet 1D.4[†]

Graphs and values of functions

Name _____ Date _____

Others in your group _____

Instructor, Teaching Assistant, and/or Recitation section _____

Directions Put first drafts of your calculations and answers on scratch paper. Take your time, work carefully, and discuss your solution with at least one other student before putting a final draft on this sheet or on other paper. Turn in all your work.

Problem 1 Generate the graph of $y = \dfrac{\sqrt{x}}{x^2+1}$ on your calculator or computer with x-min $= -1$, x-max $= 4$, x-scale $= 1$, y-min $= -0.2$, y-max $= 0.7$, and y-scale $= 0.1$. Copy the graph in Figure 1.

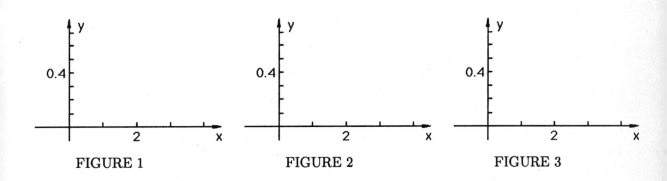

FIGURE 1 FIGURE 2 FIGURE 3

Problem 2 Find exact values of $\dfrac{\sqrt{x}}{x^2+1}$ at $x = 0.5, 2,$ and $\sqrt{2}+\sqrt{3}$ and complete the second column below. Then put ten-digit decimal values of each x and y in the third and fourth columns. Check your values by using the graph from Problem 1.

	Exact values	Ten-digit decimal values	
x	$y = \dfrac{\sqrt{x}}{x^2+1}$	$x \doteq$	$y \doteq$
0.5	$\dfrac{\sqrt{0.5}}{1.25}$	0.5	0.5656854249
2	$\tfrac{1}{5}\sqrt{2}$		
$\sqrt{2}+\sqrt{3}$			

[†]This worksheet gives practice finding values and generating graphs of functions and explores how changing a formula for a function changes its graph. See Calculator instuctions 1B.3 for tips on solving it with a calculator.

Problem 3 Generate $y = \dfrac{\sqrt{x}}{x^2+1}$ and $y = \dfrac{\sqrt{x}}{x^2+4}$ together with the ranges from Problem 1 and copy the two curves in Figure 2. How does changing 1 to 4 change the curve and why?

Problem 4 Generate $y = \dfrac{\sqrt{x}}{x^2+1}$ and $y = \dfrac{\sqrt{x-1}}{x^2+1}$ together with the ranges from Problem 1 and copy the two curves in Figure 3. How does changing $\sqrt{x}$ to $\sqrt{x-1}$ change the curve and why?

Graphing calculator workbook 31

Worksheet 1E.1†

Minimizing an area: A graphical approach

Name _____ Date _____

Others in your group _____

Instructor, Teaching Assistant, and/or Recitation section _____

Directions Put first drafts of your calculations and answers on scratch paper. Take your time, work carefully, and discuss your solution with at least one other student before putting a final draft on this sheet or on other paper. Turn in all your work.

Problem 1 You want to make a box with a square base and no top so that its volume is 200 cubic feet (Figure 1).

    a.   What would be its height h if its base is 5 feet wide?

    b.   What would be the total area A of its base and four sides if its base is 5 feet wide?

    c.   What would be its height h if its base is 10 feet wide?

    d.   What would be the total area A of its base and four sides if its base is 10 feet wide?

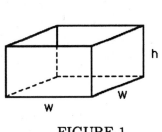

FIGURE 1

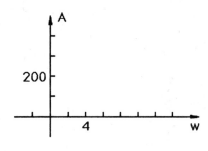

FIGURE 2

†This worksheet uses a graph generated by a graphing calculator or computer to find an approximate solution of a type of problem you will be able to solve later with calculus. See Calculator Instructions 1B.3 for tips on solving it with a calculator.

32   Graphing calculator workbook

e.   What would be its height h if its base is w feet wide?

f.   What would be the total area A of its base and four sides—expressed in terms of w—if its base is w feet wide?

g.   Generate the graph of the area function A of part (f) on your calculator or computer with x in place of w and y in place of A. Use x-min $= -4$, x-max $= 16$, x-scale $= 2$, y-min $= -100$, y-max $= 500$, and y-scale $= 100$. Copy the graph in Figure 2.

h.   Use the trace command on your calculator or the graph on your computer to find the approximate minimum total area of the base and sides of the box and the approximate width that gives that area. Do not use calculus.

Results:

Graphing calculator workbook

Worksheet 1E.2[†]

Maximizing an area: A graphical approach

Name _____ Date _____

Others in your group _____

Instructor, Teaching Assistant, and/or Recitation section _____

Directions Put first drafts of your calculations and answers on scratch paper. Take your time, work carefully, and discuss your solution with at least one other student before putting a final draft on this sheet or on other paper. Turn in all your work.

Problem 1 A rectangle with its base on the x-axis is to have its upper corners on the curve $y = \dfrac{2}{x^2 + 1}$ as in Figure 1.

    a.   What is the height of the rectangle if its lower right corner is at $x = 2$ on the x-axis?

    b.   What is its area A if its lower right corner is at $x = 2$?

    c.   What is its height if its lower right corner is at $x = 3$?

    d.   What is its area A if its lower right corner is at $x = 3$?

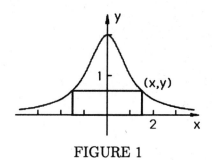

FIGURE 1

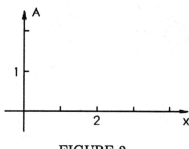

FIGURE 2

[†]This worksheet uses a graph generated by a graphing calculator or computer to find an approximate solution of a type of problem you will be able to solve later with calculus. See Calculator Instructions 1B.3 for tips on solving it with a calculator.

e. What is the height of the rectangle if its lower right corner is at x on the x-axis with an arbitrary positive number x?

f. What is the area A of the rectangle—expressed in terms of x—if its lower right corner is at x?

g. Generate the graph of the area function A of part (f) on your calculator or computer with y in place of A. Use x-min $= -0.5$, x-max $= 4.5$, x-scale $= 1$, y-min $= -0.5$, y-max $= 2.5$, and y-scale $= 1$. Copy the graph in Figure 2.

h. Use the trace command on your calculator or the graph on your computer to find the approximate maximum area of the rectangles. Then give the approximate width and height of the rectangle that has the maximum area. Do not use calculus.

Graphing calculator workbook   35

Worksheet 1E.3[†]

Maximizing the yield from an orchard: A graphical approach

Name _____ Date _____

Others in your group _____

Instructor, Teaching Assistant, and/or Recitation section _____

Directions   *Put first drafts of your calculations and answers on scratch paper. Take your time, work carefully, and discuss your solution with at least one other student before putting a final draft on this sheet or on other paper. Turn in all your work.*

Problem 1   A grower who is planning an apple orchard expects that each tree will yield 25 bushels of apples if she plants 30 trees and that for every tree over 30 that she plants, the yield for the entire orchard would drop one-fourth bushel per tree. For example, if she planted 34 trees, each tree would grow one less bushel and the yield would be $25 - 1 = 24$ bushels per tree. Then the yield from the entire orchard would be

$$\left[24 \ \frac{\text{bushels}}{\text{tree}}\right]\left[34 \ \text{trees}\right] = 816 \text{ bushels.}$$

a.   What would be the yield per tree if 38 trees are planted?

b.   What would be the yield from the entire orchard with 38 trees?

c.   What would be the yield per tree with 42 trees?

d.   What would be the yield from the entire orchard with 42 trees?

e.   What would be the yield per tree if $30 + x$ trees are planted with x any positive integer?

f.   What would be the yield from the entire orchard with $30 + x$ trees?

[†]This worksheet uses a graph generated by a graphing calculator or computer to find an approximate solution of a type of problem you will be able to solve later with calculus. See Calculator Instructions 1B.3 for tips on solving it with a calculator.

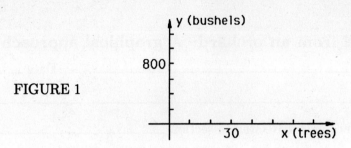

FIGURE 1

g. Generate the graph of the function from part (f) and copy it in Figure 1. Use x-min $= -10$, x-max $= 80$, x-scale $= 10$, y-min $= -200$, y-max $= 1400$, and y-scale $= 200$.

h. Use the trace command on your calculator or the graph on your computer to find the approximate maximum yield from the orchard and the approximate number of trees that would give the largest crop. Do not use calculus.

Graphing calculator workbook

Worksheet 1E.4†

Diffraction of light: A graphical approach

Name _____ Date _____

Others in your group _____

Instructor, Teaching Assistant, and/or Recitation section _____

Directions Put first drafts of your calculations and answers on scratch paper. Take your time, work carefully, and discuss your solution with at least one other student before putting a final draft on this sheet or on other paper. Turn in all your work.

A tourist in a boat is looking at a submerged lamp that is a horizontal distance of 3 meters away and 1 meter below the surface of the water (Figure 1). His eye is 1.5 meters above the surface. The light from the lamp bends when it leaves the water, and the location of the point B where it leaves the water is determined by Fermat's principle: the light takes the path that minimizes its time to go from the lamp to the man's eye. The velocity of light is 3×10^8 meters per second in air and 2.25×10^8 meters per second in water.

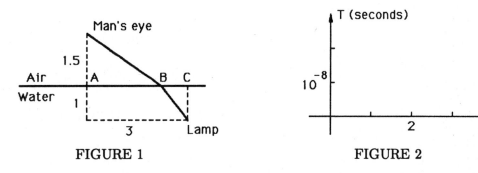

FIGURE 1   FIGURE 2

Problem 1 Suppose first that the distance $\overline{AB}$ from point A to point B is 2 meters. Then, by the Pythagorean theorem, the distance from B to the man's eye is $\sqrt{2^2 + 1.5^2} = \sqrt{6.25} = 2.5$ meters.

a. How long does it take for light to go from the point B to the man's eye? (Remember that time is distance divided by velocity.)

b. What is the distance from B to C in this case?

c. What is the distance from the lamp to B?

d. How long does it take for light to go from the lamp to point B?

---

†This worksheet uses a graph generated by a graphing calculator or computer to find an approximate solution of a type of problem you will be able to solve later with a different approach using calculus. See Calculator instructions 1B.3 for tips on solving it with a calculator.

38   Graphing calculator workbook

e.   How long does it take for light to go from the lamp to the man's eye?

Problem 2   Next suppose that the distance $\overline{AB}$ from point A to point B is x meters, where x is an arbitrary number between 0 and 3. Then the distance from B to the man's eye is $\sqrt{x^2 + 1.5^2} = \sqrt{x^2 + 2.25}$ meters.

a.   How long does it take for light to go from B to the man's eye?

b.   What is the distance from B to C?

c.   What is the distance from the lamp to B?

d.   How long does it take for light to go from the lamp to B?

e.   Give the amount of time T it takes for light to go from the lamp to the man's eye as a function of x.

f.   Generate the graph of the function T of part (e) on your calculator or computer. Use y in place of T, x-min $= -0.5$, x-max $= 4$, x-scale $= 1$, y-min $= -0.5 \times 10^{-8}$, y-max $= 3 \times 10^{-8}$, and y-scale $= 1 \times 10^{-8}$. Copy the graph in Figure 2.

g.   Use the trace command on your calculator or the graph on your computer to find the approximate minimum time for the light to go from the lamp to the man's eye and the approximate distance from the point A to where the light ray leaves the water. Do not use calculus.

Graphing calculator workbook

Worksheet 1F.1[†]

Solving equations by factoring

Name _____ Date _____

Others in your group _____

Instructor, Teaching Assistant, and/or Recitation section _____

Directions   Put first drafts of your calculations and answers on scratch paper. Take your time, work carefully, and discuss your solution with at least one other student before putting a final draft on this sheet or on other paper. Turn in all your work.

Problem 1   Find all solutions x of the equation $x^4 - 5x^2 + 4 = 0$ by first factoring $x^4 - 5x^2 + 4$ with x^2 as the variable. Show your steps. Generate the curve $y = x^4 - 5x^2 + 4$ on your calculator or computer to guide your reasoning, if necessary, and to check your answers. Copy the graph in Figure 1. Use x-min $= -2.5$, x-max $= 2.5$, x-scale $= 1$, y-min $= -4$, y-max $= 8$, and y-scale $= 1$.

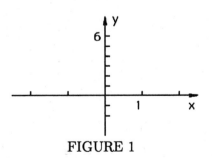

FIGURE 1

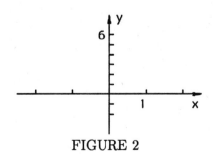

FIGURE 2

[†]This worksheet shows how a graphing calculator or computer can be used to check answers obtained by algebra.

Problem 2 Find all solutions x of the equation $x^4 - 5x^2 + 4 = 4$. Show your steps. Generate the curve $y = x^4 - 5x^2 + 4$ and the line $y = 4$ on your calculator or computer to guide your reasoning and check your answers. Give exact answers, not approximate decimal values. Copy the graphs in Figure 2. Use x-min $= -2.5$, x-max $= 2.5$, x-scale $= 1$, y-min $= -4$, y-max $= 8$, and y-scale $= 1$.

Worksheet 1F.2†

Solving equations with the quadratic formula

Name _____ Date _____

Others in your group _____

Instructor, Teaching Assistant, and/or Recitation section _____

Directions *Put first drafts of your calculations and answers on scratch paper. Take your time, work carefully, and discuss your solution with at least one other student before putting a final draft on this sheet or on other paper. Turn in all your work.*

Problem 1 Find all solutions x of the equation $x^2 - 4x = 4$. Show your steps and give exact answers, not approximate decimal values. Generate the curve $y = x^2 - 4x$ and the line $y = 4$ on your calculator or computer with x-min $= -3$, x-max $= 7$, x-scale $= 1$, y-min $= -5$, y-max $= 10$, and y-scale $= 1$ to guide your reasoning, if necessary, and to check your answers. Copy the graphs in Figure 1.

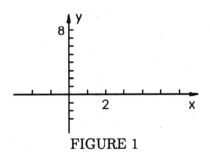

FIGURE 1

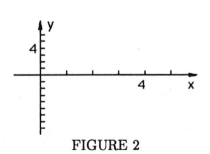

FIGURE 2

---

†This worksheet shows how a graphing calculator or computer can be used to check answers obtained by algebra.

Problem 2   Find all solutions x of the equation $x^3 - 7x^2 + 10x = -x$. Show your steps and give exact answers, not approximate decimal values. Generate the curve $y = x^3 - 7x^2 + 10x$ and the line $y = -x$ on your calculator or computer to guide your reasoning, if necessary, and to check your answers. Use x-min $= -1$, x-max $= 6$, x-scale $= 1$, y-min $= -10$, y-max $= 8$, and y-scale $= 1$. Copy the graphs in Figure 2.

Graphing calculator workbook

Worksheet 1F.3[†]

Two equations with absolute values and an inequality

Name _____ Date _____

Others in your group _____

Instructor, Teaching Assistant, and/or Recitation section _____

Directions Put first drafts of your calculations and answers on scratch paper. Take your time, work carefully, and discuss your solution with at least one other student before putting a final draft on this sheet or on other paper. Turn in all your work.

Problem 1 Solve $|x^2 - 4| = 2$ for x. Justify your conclusion. Remember that for a nonnegative number k, the equation $|y| = k$ is equivalent to the condition $y = \pm k$ Generate the curve $y = |x^2 - 4|$ and the line $y = 2$ on your calculator or computer to guide your reasoning and check your answers. Use x-min $= -3.5$, x-max $= 3.5$, x-scale $= 1$, y-min $= -1$, y-max $= 5$, and y-scale $= 1$. Copy the graphs in Figure 1.

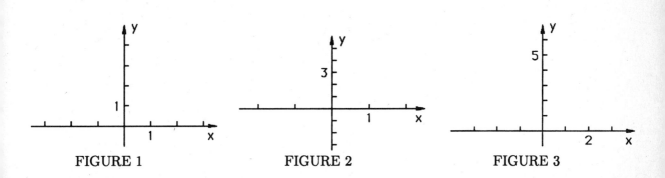

FIGURE 1   FIGURE 2   FIGURE 3

[†]This worksheet shows how you can use a graphing calculator or computer to help you find all solutions of equations and inequalities.

44   Graphing calculator workbook

Problem 2   Solve $x^4 - 3x^2 \geq 0$. Justify your answer. Generate the curve $y = x^4 - 3x^2$ on your
calculator or computer to guide your reasoning. Copy the graph in Figure 2. Use
x-min $= -2.5$, x-max $= 2.5$, x-scale $= 1$, y-min $= -4$, y-max $= 6$, and y-scale $= 1$.

Problem 3   Solve $|x + 1| + |x| = 3.4$. Justify your answer. Remember that $|x|$ equals x for
$x \geq 0$ and equals $-x$ for $x < 0$ and that $|x + 1|$ equals $x + 1$ for $x \geq -1$ and equals
$-x - 1$ for $x < -1$. Generate $y = |x + 1| + |x|$ and $y = 3.4$ on your calculator or
computer to guide your reasoning. Copy the graphs in Figure 3. Use x-min $= -4$,
x-max $= 4$, x-scale $= 1$, y-min $= -1$, y-max $= 6$, and y-scale $= 1$.

Graphing calculator workbook   45

Worksheet 1F.4[†]

Domains of functions involving square roots

Name _____ Date _____

Others in your group _____

Instructor, Teaching Assistant, and/or Recitation section _____

Directions   Put your initial calculations and first drafts of your conclusions on scratch paper. Take your time, work carefully, and give your conclusions in well written sentences. Discuss your solution with at least one other student before putting a final draft on this sheet or other sheets of paper, as needed. Turn in all your work.

Problem 1   Find the domain of $\sqrt{1+x^3}$. Justify your conclusion. Generate the curve $y = \sqrt{1+x^3}$ on your calculator or computer to guide your reasoning. Use x-min $= -2$, x-max $= 2$, x-scale $= 1$, y-min $= -1$, y-max $= 3$, and y-scale $= 1$. Copy the graph in Figure 1.

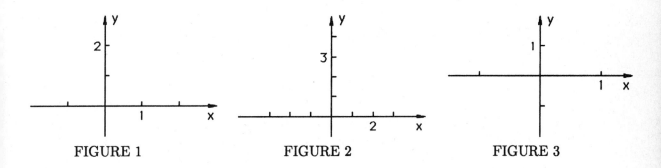

FIGURE 1   FIGURE 2   FIGURE 3

The domain is:

Justification:

[†]This worksheet shows how a graphing calculator or computer can help you find domains of functions.

46   Graphing calculator workbook

Problem 2   Find the domain of $\sqrt{1+x^2}$. Justify your conclusion. Generate the curve $y = \sqrt{1+x^2}$ on your calculator or computer to guide your reasoning. Use x-min $= -4.5$, x-max $= 4.5$, x-scale $= 1$, y-min $= -1$, y-max $= 5$, and y-scale $= 1$. Copy the graph in Figure 2.

The domain is:

Justification:

Problem 3   Find the domain of $\sqrt{1+x} - \sqrt{1-x}$. Justify your conclusion. Generate the curve $y = \sqrt{1+x} - \sqrt{1-x}$ on your calculator or computer to guide your reasoning. Use x-min $= -1.5$, x-max $= 1.5$, x-scale $= 1$, y-min $= -2$, y-max $= 2$, and y-scale $= 1$. Copy the graph in Figure 3.

The domain is:

Justification:

Graphing calculator workbook

Worksheet 1G.1[†]

Two-sided finite limits: Geometric and numerical investigations

Name _____ Date _____

Others in your group _____

Instructor, Teaching Assistant, and/or Recitation section _____

Directions Put first drafts of your calculations and answers on scratch paper. Take your time, work carefully, and discuss your solution with at least one other student before putting a final draft on this sheet or on other paper. Turn in all your work.

Problem 1a Generate the graph of $f(x) = \dfrac{x^4 - 1}{x^2 - 1}$ on your calculator or computer for $-2 \leq x \leq 3$, $-2 \leq y \leq 10$ and copy it in Figure 1. Then use the trace procedure on a calculator or the graph on a computer to predict $\lim\limits_{x \to 1} f(x)$.

b. Generate the graph of the function from part (a) with $0.98 \leq x \leq 1.03$, x-scale $= 0.01$ and the range of y from part (a) and copy it in Figure 2. Use the trace procedure on a calculator or the graph on a computer to support or improve your prediction of $\lim\limits_{x \to 1} f(x)$.

c. Generate the graph with $0.9998 \leq x \leq 1.0003$, x-scale $= 0.0001$, and the same range of y and copy it in Figure 3. Use the trace procedure on a calculator or the graph on a computer to support or improve your prediction of $\lim\limits_{x \to 1} f(x)$.

d. Support or improve your prediction of the limit by completing the table on the back with values of $f(x)$.

e. Find the exact limit in part (a) by factoring and cancellation.

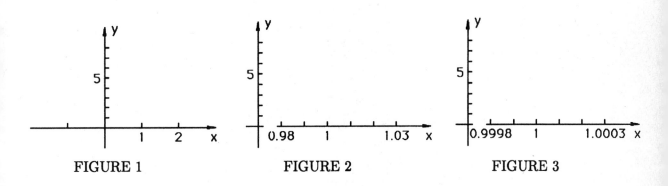

FIGURE 1   FIGURE 2   FIGURE 3

[†]Problem 1 calls for predicting a two-sided finite limit from graphs and approximate calculations and then finding its exact value by factoring and cancellation. In Problem 2, graphs are used to study how close the value of the function $f(x)$ is to its limit for x near its limiting value. See Calculator instructions 1B.3 for tips on solving these problems with a calculator.

x	$f(x)$
0.9	
0.999	
0.999999	
1.1	
1.001	
1.000001	

Problem 2a Generate the graph of the function $f(x)$ from Problem 1 with the lines $y = 1.5$, $y = 2$ and $y = 2.5$ for $-2 \leq x \leq 3$, $0.5 \leq y \leq 3$ and copy them in Figure 4.

b. Because $f(x) \to 2$ as $x \to 1$, $f(x)$ satisfies $1.5 < f(x) < 2.5$ for x sufficiently close to 1 and $\neq 1$. Use the graph to determine whether $1.5 < f(x) < 2.5$ for all x with $0.8 < x < 1$.

c. Is $1.5 < f(x) < 2.5$ for all x with $1 < x < 1.3$?

d. Generate the graph of the function $f(x)$ from Problem 1 with the lines $y = 1.999$, $y = 2$, and $y = 2.001$. Use $0.998 \leq x \leq 1.003$, x-scale $= 0.001$, $1.997 \leq y \leq 2.003$, and y-scale $= 0.001$. Copy the curve and three lines in Figure 5.

e. Because $f(x) \to 2$ as $x \to 1$, $f(x)$ satisfies $1.999 < f(x) < 2.001$ for x sufficiently close to 1 and $\neq 1$. Use the trace on a calculator or the graph on a computer to find numbers $a < 1$ and $b > 1$ such that $1.999 < f(x) < 2.001$ for all x with $a < x < 1$ and for all x with $1 < x < b$.

f. Figures 3 and 5 have the same range of x, but the graph of $f(x)$ in Figure 3 looks like a horizontal line, while the graph of $f(x)$ in Figure 5 does not appear horizontal. Explain.

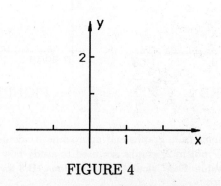

FIGURE 4

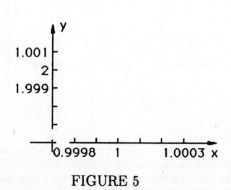

FIGURE 5

Graphing calculator workbook

Worksheet 1G.2[†]

One-sided finite limits: Geometric and numerical investigations

Name _____ Date _____

Others in your group _____

Instructor, Teaching Assistant, and/or Recitation section _____

Directions Put first drafts of your calculations and answers on scratch paper. Take your time, work carefully, and discuss your solution with at least one other student before putting a final draft on this sheet or on other paper. Turn in all your work.

Problem 1 Generate the graph of $P(x) = \dfrac{x-2}{|x-2|}$ on your calculator or computer for $-1 \leq x \leq 4, -1.5 \leq y \leq 1.5$ and copy it in Figure 1. Use the trace command on a calculator or the graph on a computer to predict the values of each of the following limits.

a. $\lim\limits_{x \to 2^+} P(x)$:

b. $\lim\limits_{x \to 2^-} P(x)$:

c. $\lim\limits_{x \to 0^+} P(x)$:

d. $\lim\limits_{x \to 3^-} P(x)$:

e.   Explain why your predictions are correct.

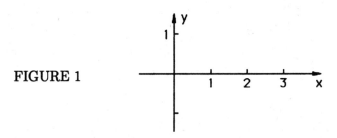

FIGURE 1

[†]This worksheet calls for predicting one-sided finite limits from graphs and then finding their exact values from the formulas. See Calculator instructions 1B.3 for tips on solving it with a calculator.

Problem 2 Generate the graph of

$$Q(x) = \frac{\sqrt{x^2 - 1}}{\sqrt{x - 1}}$$

for $-1 \leq x \leq 9, -1 \leq y \leq 4$ and copy it in Figure 2.

a. Use the trace command on a calculator or the graph on a computer to predict
$$\lim_{x \to 1^+} Q(x).$$

b. Generate the graph of $Q(x)$ with $0.998 \leq x \leq 1.003$, x-scale $= 0.001$, $-1 \leq y \leq 4$, and copy it in Figure 3. Use the graph to support or improve your prediction of the limit in part (a).

c. Find the exact value of the limit by factoring and calcellation.

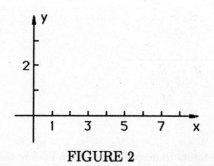

FIGURE 2

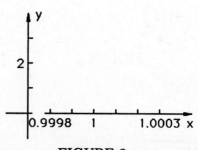

FIGURE 3

Graphing calculator workbook

Worksheet 1G.3[†]

Infinite limits: Geometric and numerical investigations

Name _____ Date _____

Others in your group _____

Instructor, Teaching Assistant, and/or Recitation section _____

Directions Put first drafts of your calculations and answers on scratch paper. Take your time, work carefully, and discuss your solution with at least one other student before putting a final draft on this sheet or on other paper. Turn in all your work.

Problem 1 Sketch the graph of $A(x) = \begin{cases} -3 - \dfrac{3}{x} & \text{for } x < 0 \\ 3 - \dfrac{3}{x} & \text{for } x > 0 \end{cases}$

in Figure 1 without using a calculator or computer. Then complete the table of decimal values of $A(x)$ on the next page. Use the table, the graph, and the formulas for $A(x)$ to predict the limits of $A(x)$ as

a. $x \to -\infty$:

b. $x \to 0^-$:

c. $x \to 0^+$:

d. $x \to 0$:

e. $x \to \infty$:

f. What are the vertical and horizontal asymptotes of the graph?

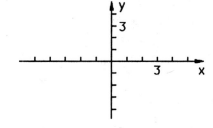

FIGURE 1

[†]This worksheet deals with examples of infinite limits. See Calculator instuctions 1B.3 for tips on solving it with a calculator.

x	$A(x) \doteq$
-3	-2
-100	2.97
-1×10^6	
-0.1	
-0.001	
-1×10^{-6}	2999997

x	$A(x) \doteq$
0.1	-27
0.001	
1×10^{-6}	
3	
100	
1×10^6	2.999997

Problem 2a $x^{2/3}$ tends to ∞ as $x \to \infty$. Consequently, $x^{2/3}$ is greater than 10 for x sufficiently large. Generate the graph of $x^{2/3}$ and the line $y = 10$ on your calculator or computer with $-5 \leq x \leq 50$, x-scale $= 10$, $-1.5 \leq y \leq 15$ and y-scale $= 5$ and copy them in Figure 2. Is $x^{2/3} > 10$ for all $x > 35$?

    b.   Generate $y = x^{2/3}$ and $y = 500$ with $-2000 \leq x \leq 20000$, x-scale $= 5000$, $-80 \leq y \leq 800$, y-scale $= 100$. Copy them in Figure 3. $x^{2/3}$ is also greater than 500 for x sufficiently large. Is $x^{2/3} > 500$ for all $x > 10,000$?

    c.   Generate $y = x^{2/3}$ and $y = 1000$ with $-5000 \leq x \leq 50000$, x-scale $= 10000$, $-150 \leq y \leq 1500$, y-scale $= 500$. Copy them in Figure 4. Because $x^{2/3} \to \infty$ as $x \to \infty$, there is a constant a such that $x^{2/3} > 1000$ for all $x > a$. Find such a number a.

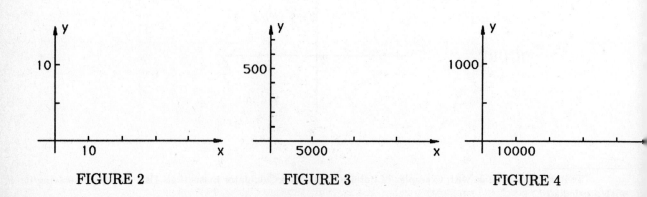

FIGURE 2   FIGURE 3   FIGURE 4

Graphing calculator workbook   53

Worksheet 1G.4[†]

Continuity and the Extreme and Intermediate value theorems

Name _____ Date _____

Others in your group _____

Instructor, Teaching Assistant, and/or Recitation section _____

Directions Put your initial calculations and first drafts of your conclusions on scratch paper. Take your time, work carefully, and give your conclusions in well written sentences. Discuss your solution with at least one other student before putting a final draft on this sheet or other sheets of paper, as needed. Turn in all your work.

Problem 1 Generate the graph of $\dfrac{x^3 - 1}{x - 1}$ on your calculator or computer for $-2.5 \leq x \leq 2.5$, $-2 \leq y \leq 8$ and copy it in Figure 1. You cannot tell from the graph whether or not $\dfrac{x^3 - 1}{x - 1}$ is continuous in the interval $[-2, 2]$. Is it? Explain.

Problem 2 Generate the graph of $\sqrt{x}$ on your calculator or computer for $-2 \leq x \leq 15$, $-1 \leq y \leq 4$ and copy it in Figure 2. You cannot tell from the graph whether or not $\sqrt{x}$ is continuous in the interval $[0, 10]$. Is it? Explain.

Problem 3 Figure 3 shows the graph of

$$f(x) = \begin{cases} x^2 & \text{for } x \leq 1 \\ x^2 - 2x - 1 & \text{for } 1 < x < 3 \\ 4 & \text{for } x \geq 3. \end{cases}$$

a. Find a number b such that $f(x)$ is continuous in $(-\infty, b]$ but not in $(-\infty, b+1)$.

b. Find numbers a and b such that $f(x)$ is continuous in (a, b) but not in $(a, b]$.

c. Find the least number a such that $f(x)$ is continuous in $[a, \infty)$.

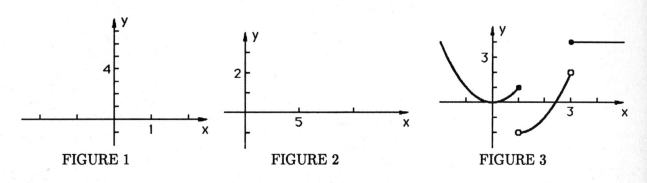

FIGURE 1 FIGURE 2 FIGURE 3

[†]This worksheeet uses the defintion of continuity in intervals, expressed in terms of one-sided limits, and explores the hypotheses and conclusions of the Extreme and Intermediate value theorems.

54   Graphing calculator workbook

d. Find numbers a and b such that $f(x)$ is continuous in $[a, b]$ and has a maximum
    and a minimum value in $[a, b]$. What are the maximum and minimum values? Are
    the hypotheses of the Extreme value theorem satisfied in this example? Is the
    conclusion satisfied?

e. Find numbers a and b such that $f(x)$ is discontinuous in $[a, b]$ and has a maximum
    and a minimum value in $[a, b]$. What are the maximum and minimum? Are the
    hypotheses of the Extreme value theorem satisfied in this example? Is the conclusion
    satisfied?

f. Find numbers a and b such that $f(x)$ is continuous in (a, b) and does not have a
    maximum value in (a, b). Are the hypotheses of the Extreme value theorem satisfied
    in this example? Is the conclusion satisfied?

g. Find numbers a, b, and L such that $f(x)$ is continuous in $[a, b]$, $f(a) \leq L \leq f(b)$
    and the equation $f(x) = L$ has a solution in $[a, b]$. What are the solutions? Are
    the hypotheses of the Intermediate value theorem satisfied in this example? Is the
    conclusion satisfied?

h. Find numbers a, b, and L such that $f(x)$ is discontinuous in $[a, b]$, $f(a) \leq L \leq f(b)$
    and the equation $f(x) = L$ has a solution in $[a, b]$. Give a solution. Are the
    hypotheses of the Intermediate value theorem satisfied in this example? Is the
    conclusion satisfied?

i. Find numbers a, b, and L such that $f(x)$ is discontinuous in $[a, b]$, $f(a) \leq L \leq f(b)$
    and the equation $f(x) = L$ does not have a solution in $[a, b]$. Are the hypotheses of
    the Intermediate value theorem satisfied in this example? Is the conclusion satisfied?

Graphing calculator workbook

Worksheet 1H.1†

A ferris wheel and the Berlin wall

Name _____ Date _____

Others in your group _____

Instructor, Teaching Assistant, and/or Recitation section _____

Directions Put first drafts of your calculations and answers on scratch paper. Take your time, work carefully, and discuss your solution with at least one other student before putting a final draft on this sheet or on other paper. Turn in all your work.

Problem 1 The largest of all ferris wheels was built by G. W. Gale Ferris for the 1893 World's Columbian Exposition in Chicago. It was 250 feet in diameter and could carry 2160 passengers. Suppose that it turned counterclockwise as viewed in Figure 1, that its center was 150 feet above the ground, that it took eight minutes for a full revolution, and that the car indicated by the dot was 150 feet above the ground and rising at $t = 0$ (minutes).

a. Which of the curves in Figures 2 through 6 is the graph of the height y of the car above the ground as a function of t? Justify your choice without finding a formula for $y(t)$.

b. Let x (radians) denote the angle through which the wheel has turned from time 0 to time $t \geq 0$. What is x at $t = 0$, $t = 2$, $t = 4$, $t = 6$, and $t = 8$?

c. Give a formula for the angle x as a function of t.

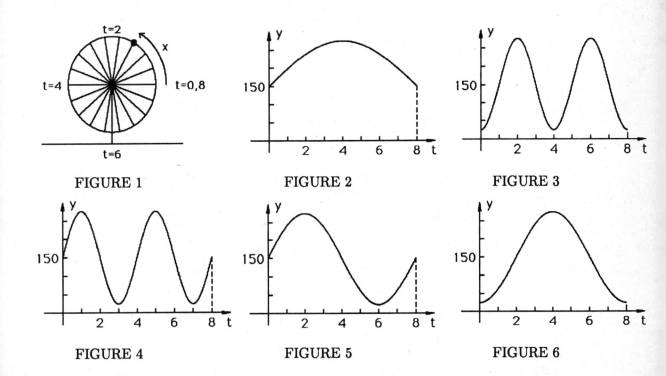

†This worksheet is a partial review of the definitions of the trigonometric functions and their graphs.

d. Use a trigonometric function to give a formula for y as a function of t. Check your result by generating the graph on your calculator or computer.

Problem 2 It is 1965. A searchlight shining on the Berlin Wall from East Berlin is 50 meters from the wall and rotates one revolution every minute, counterclockwise as viewed from above (Figure 7). Two beams from the searchlight shine parallel to the ground in opposite directions, so light sweeps along the wall every half minute.

a. Suppose that v measures the distance (meters) along the wall with $v = 0$ at the point opposite the searchlight, with positive v for one direction and with negative v for the other. Also suppose that at $t = 0$ (minutes) one beam is shining directly at the point on the wall where $v = 0$. Which of the curves in Figures 8 through 10 is the graph of v as a function of t? Explain without finding a formula for $v(t)$.

b. Suppose that x is the angle in Figure 7 measured in radians with $x = 0$ radians at $t = 0$ minutes. Explain why $x = 2\pi t$.

c. Use a trigonometric function to give a formula for v as a function of t. Check by generating the graph of the function on your calculator or computer for $-0.25 \leq t \leq 1.25, -150 \leq v \leq 150$.

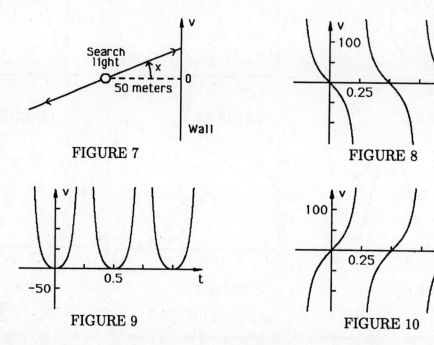

FIGURE 7

FIGURE 8

FIGURE 9

FIGURE 10

Worksheet 1H.2†

Equations involving trigonometric functions

Name _____ Date _____

Others in your group _____

Instructor, Teaching Assistant, and/or Recitation section _____

Directions   Put first drafts of your calculations and answers on scratch paper. Take your time, work carefully, and discuss your solution with at least one other student before putting a final draft on this sheet or on other paper. Turn in all your work.

Problem 1   Find all values of x such that **(a)** $\sin x = 0$ and **(b)** $\sin x = 1$. Sketch the line $y = 1$ and the curve $y = \sin x$ in the xy-plane of Figure 1 for $-2\pi \leq x \leq 2\pi$ without using a calculator or computer, and use this sketch to help you find the solutions for $-2\pi \leq x \leq 2\pi$. Then use the fact that $\sin x$ is periodic of period 2π to find all the other solutions.

Problem 2   Find all solutions x of $\sin x = \sin^2 x$, by first factoring $\sin^2 x - \sin x = 0$ with $\sin x$ as the variable and solving for $\sin x$. Generate the curves $y = \sin x$ and $y = \sin^2 x$ together on your calculator or computer with $-7 \leq x \leq 7$, x-scale $= 1.57 \doteq \pi/2$, and $-1.5 \leq y \leq 1.5$ to guide your reasoning. (Enter $\sin^2 x$ as $(\sin x)^2$.) Copy the curves in Figure 2. Why does replacing $\sin x$ by $\sin^2 x$ change the shape of the curve as it does?

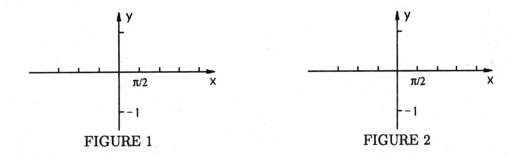

FIGURE 1   FIGURE 2

---

†These problems review properties of trigonometric functions.

Problem 3 Find all solutions x of $\tan^3 x = 3\tan x$. Generate the curves $y = \tan^3 x$ and $y = 3\tan x$ together on your calculator or computer with $-1.3 \leq x \leq 1.3$, x-scale $= 0.52 \doteq \pi/6$, $-10 \leq y \leq 10$, and y-scale $= 2$ to help you find the solutions for $-\frac{1}{2}\pi \leq x \leq \frac{1}{2}\pi$. Copy the curves in Figure 3. Then use the fact that $\tan x$ is periodic of period π to find all the other solutions.

Problem 4 Find all solutions x of $2\sec x = \sec^3 x$, by first factoring $\sec^3 x - \sec x = 0$ with $\sec x$ as variable and solving for $\sec x$ and then finding the values of $\cos x$. Generate the curves $y = 2\sec x$ ($= 2/\cos x$) and $y = \sec^3 x$ ($= 1/(\cos x)^3$) together on your calculator or computer with $-1.3 \leq x \leq 1.3$, x-scale $= 1.05 \doteq \pi/3$, and $-1 \leq y \leq 6$ to guide your reasoning. Copy the curves in Figure 4.

Problem 5 Show that the equation $\tan x = -\sec x$ has no solutions. Generate the curves $y = \tan x$ and $y = -\sec x$ with $-1.5 \leq x \leq 1.5$, x-scale $= 0.785 \doteq \pi/4$, and $-5 \leq y \leq 5$ on your calculator or computer to guide your reasoning. Copy the curves in Figure 5.

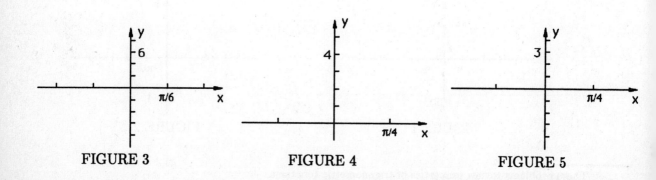

FIGURE 3   FIGURE 4   FIGURE 5

Graphing calculator workbook

Worksheet 1H.3[†]

Limits involving trigonometric functions

Name _____ Date _____

Others in your group _____

Instructor, Teaching Assistant, and/or Recitation section _____

Directions Put your initial calculations and first drafts of your conclusions on scratch paper. Take your time, work carefully, and give your conclusions in well written sentences. Discuss your solution with at least one other student before putting a final draft on this sheet or other sheets of paper, as needed. Turn in all your work.

Problem 1 Generate the graph of $\tan^2 x$ on your calculator or computer and copy it in Figure 1. Use $-1.57 \leq x \leq 4.71$, x-scale $= 1.57 \doteq \pi/2$, $-5 \leq y \leq 25$, and y-scale $= 5$.

  a. Use the graph to predict $\lim\limits_{x \to \pi/2} \tan^2 x$:

  b. Use properties of the tangent function to justify your conclusion:

Problem 2 Generate the graph of $x + \sin x$ and copy it in Figure 2. Use $-5 \leq x \leq 20$, x-scale $= 5$, $-5 \leq y \leq 25$, and y-scale $= 5$.

  a. Use the graph to predict $\lim\limits_{x \to \infty} (x + \sin x)$:

  b. Use a property of the sine function to justify your conclusion:

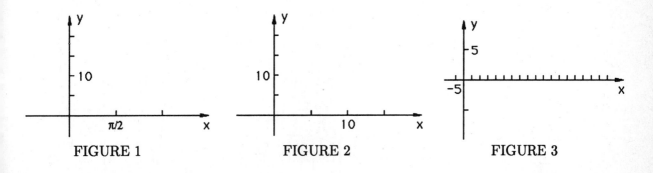

FIGURE 1   FIGURE 2   FIGURE 3

[†]These problems review properties of the trigonometric functions.

Problem 3 Generate the graph of $\sqrt{x}\sin x$ and copy it in Figure 3. Use $-10 \le x \le 100$, $-10 \le y \le 10$, x-scale $= 5$, and y-scale $= 5$. What can you say about $\lim\limits_{x \to \infty}(\sqrt{x}\sin x)$?

Problem 4 Generate the graph of $f(x) = \dfrac{\sin(\pi x)}{|x-1|}$ for $-1 \le x \le 4$, $-4 \le y \le 4$ and copy it in Figure 4.

a. Use the trace command on your calculator or the graph on your computer to predict approximate decimal values of $\lim\limits_{x \to 1^-} f(x)$ and $\lim\limits_{x \to 1^+} f(x)$.

b. Guess the exact values of the limits if you can.

Problem 5 Figures 5 and 6 show the graphs of $\sin(1/x)$ and $x\sin(1/x)$. (These graphs are not very clear on a calculator because of poor resolution.) What can you say about their limits as $x \to 0$? Explain.

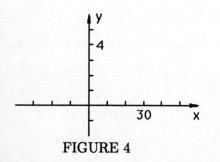

FIGURE 4

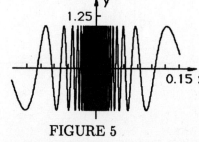

FIGURE 5

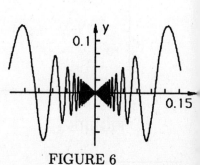

FIGURE 6

Graphing calculator workbook

Worksheet 1I.1†

Graphs involving logarithms and exponential functions

Name _____ Date _____

Others in your group _____

Instructor, Teaching Assistant, and/or Recitation section _____

Directions Put first drafts of your calculations and answers on scratch paper. Take your time, work carefully, and discuss your solution with at least one other student before putting a final draft on this sheet or on other paper. Turn in all your work.

Recall that for constants $b > 1$, the logarithm to the base b is denoted $\log_b x$ and is the inverse of b^x. For example, $\log_2(8)$ equals 3 because $2^3 = 8$. Because b^x and $\log_b x$ are inverse functions, their graphs are mirror images about the line $y = x$ in an xy-plane with equal scales on the axes.

Problem 1 Figure 1 shows the graphs of 2^x, 3^x, $\log_2 x$, and $\log_3 x$. Match the functions to their graphs without using a calculator or computer and explain how you determine which graph is which. Check by generating the curves on your calculator or computer for $-5 \leq x \leq 4$, $-2 \leq y \leq 4$. ($\log_{10} x$ is probably given as $\log x$ and then $\log_b x$ can be obtained as $(\log x) \div (\log b)$.)

Problem 2 Use properties of exponential functions to match the following functions to their graphs in Figures 2 through 6. Explain your reasoning. Check your results by generating the graphs on your calculator or computer for $-2 \leq x \leq 2$, $-1 \leq y \leq 6$.

(a) 2^{-x}, (b) $2 + 2^x$, (c) 2^{1+x}, (d) $2^{1/x}$, (e) $1 + x + 1^x$

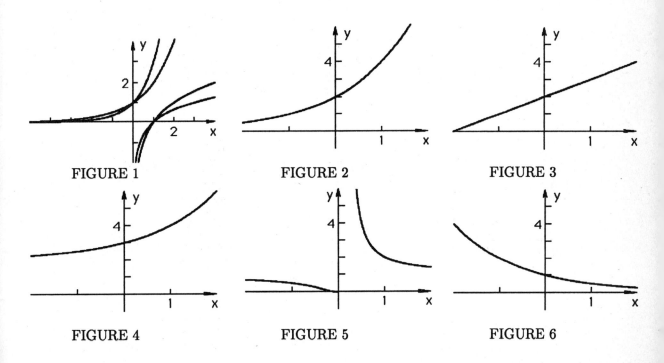

FIGURE 1   FIGURE 2   FIGURE 3

FIGURE 4   FIGURE 5   FIGURE 6

---

†These problems review properties of logarithms and exponential functions to the base b.

Problem 3 Use properties of exponential functions to match the following functions to their graphs in Figures 7 through 9. Explain your reasoning. Check your results by generating the graphs on your calculator or computer with the ranges of x and y in the figures.

(a) 2^{x^2}, (b) 2^{2-x^2}, (c) $5 - 2^{x^2}$

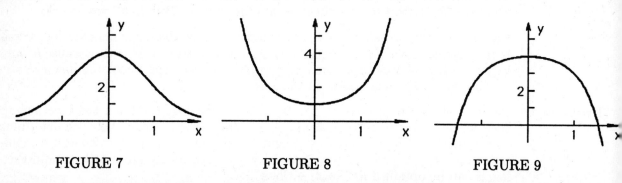

FIGURE 7　　　　　　　FIGURE 8　　　　　　　FIGURE 9

Problem 4 Use properties of logarithms to match the following functions to their graphs in Figures 10 through 12. Explain your reasoning. Check your results by generating the graphs on your calculator or computer.

(a) $x \log_2 x$, (b) $(\log_2 x)^2$, (c) $\dfrac{1}{\log_2 x}$

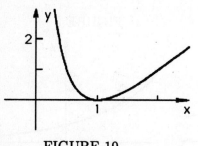 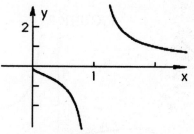 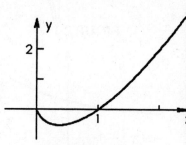

FIGURE 10　　　　　　　FIGURE 11　　　　　　　FIGURE 12

Graphing calculator workbook

Worksheet 1I.2[†]

Equations involving logarithms and exponential functions

Name _____ Date _____

Others in your group _____

Instructor, Teaching Assistant, and/or Recitation section _____

Directions Put first drafts of your calculations and answers on scratch paper. Take your time, work carefully, and discuss your solution with at least one other student before putting a final draft on this sheet or on other paper. Turn in all your work.

Problem 1 Use properties of exponential functions to find all solutions x of $2^x = 8(2^{-x})$. Give your reasoning. Check by generating the curves $y = 2^x$ and $y = 8(2^{-x})$. Use $-1 \leq x \leq 4$ and pick a suitable range of y by trial and error. Copy the graphs and put a scale on the y-axis in Figure 1.

Problem 2 Solve $4^{x+1} = 4^x + 6$ for x. Give your reasoning. Check your answer(s) by generating the curves $y = 4^{x+1}$ and $y = 4^x + 6$ together on your calculator or computer. Use $-0.5 \leq x \leq 1.5$ and pick a suitable y-range by trial and error. Copy the curves in Figure 2 and put a scale on the y-axis.

Problem 3 Solve $3^{2x-x^2} = 3$ for x. Justify your answer(s). Check by generating the curve $y = 3^{2x-x^2}$ and the line $y = 3$ together on your calculator or computer. Use $-1 \leq x \leq 4$ and pick a suitable y-range by trial and error. Copy the graphs in Figure 3 and put a scale on the y-axis.

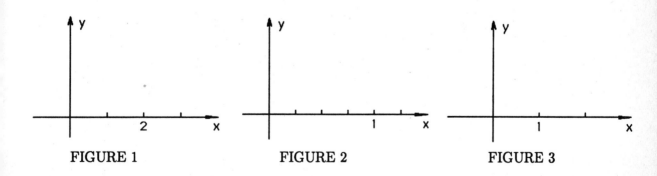

FIGURE 1 FIGURE 2 FIGURE 3

[†]These problems review properties of logarithms and exponential functions to the base b. See Calculator instructions 1B.3 for tips on solving these problems with a calculator.

Problem 4 Use properties of exponential functions to find all solutions x of $(2^x)^2 = 2^{x^2}$. Justify your answer(s). Check by generating the curves $y = (2^x)^2$ and $y = 2^{x^2}$. Use $-1 \leq x \leq 2.5$ and a suitable range for y. Copy the graphs and put a scale on the y-axis in Figure 4.

Problem 5 Use properties of logarithms to solve $\log_{10}(x^3) = 8 - \log_{10} x$ for x. Check your answer(s) by generating the curves $y = \log_{10}(x^3)$ and $y = 8 - \log_{10} x$ on the same screen of your calculator or computer. Use $-50 \leq x \leq 300$, x-scale $= 100$, and a suitable y-range. Copy the curves in Figure 5 and put a scale on the y-axis.

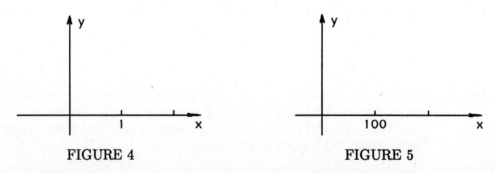

FIGURE 4   FIGURE 5

Problem 6 Solve $(\log_{10} x)^2 = 4$. Give your reasoning and check your answer(s) by generating the curve $y = (\log_{10} x)^2$ and the line $y = 4$ with $-25 \leq x \leq 150$ to show the large solution and with $-0.05 \leq x \leq 0.2$ to show the small solution. Pick a suitable y-range and copy the curves in Figures 6 and 7. Put scales on the y-axes.

Problem 7 Figure 8 shows the graph of the function $y(x)$ such that (i) $\log_{10} y = -x^2 + C$ with a constant C and (ii) $y(0) = 10$. Use (i) and (ii) to find the value of C and then solve for y. Check by generating the graph of this function on your calculator or computer with $-1.5 \leq x \leq 1.5, -3 \leq y \leq 15$.

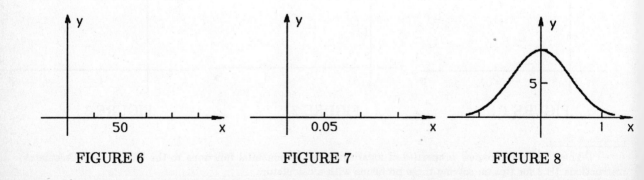

FIGURE 6   FIGURE 7   FIGURE 8

Graphing calculator workbook

Worksheet 1I.3†

The algebra of logarithms and exponential functions

Name _____ Date _____

Others in your group _____

Instructor, Teaching Assistant, and/or Recitation section _____

Directions   *Put first drafts of your calculations and answers on scratch paper. Take your time, work carefully, and discuss your solution with at least one other student before putting a final draft on this sheet or on other paper. Turn in all your work.*

Problem 1   Use properties or values of exponential functions to determine whether the functions $(2^x + 2^{-x})^2$ and $4^x + 2 + \dfrac{1}{4^x}$ are equal for all x. Give your reasoning. Check by generating the curves $y = (2^x + 2^{-x})^2$ and $y = 4^x + 2 + \dfrac{1}{4^x}$ together on your calculator or computer. Use $-2 \leq x \leq 2$ and pick a suitable range of y by trial and error. Copy the graphs and put the scale on the y-axis in Figure 1.

Problem 2   Use properties or values of logarithms to determine whether $\log_{10}(x^3 + 7)$ equals $3\log_{10} x + \log_{10}(7)$ for all x where both are defined. Give your reasoning. Check by generating the curves $y = \log_{10}(x^3 + 7)$ and $y = 3\log_{10} x + \log_{10}(7)$ together on your calculator or computer. Use $-1 \leq x \leq 3$ and a suitable y-range. Copy the curves in Figure 2 and put the scale on the y-axis.

Problem 3   Use properties or values of exponential functions to determine whether $(2^{1+x})^2$ equals 2^{1+2x+x^2} for all x. Justify your answer. Check by generating $y = (2^{1+x})^2$ and $y = 2^{1+2x+x^2}$ together on your calculator or computer. Use $-0.5 \leq x \leq 1.5$ and a suitable y-range. Copy the curves in Figure 3 and put the scale on the y-axis.

Problem 4   Are $\sqrt{2^{6x}}$ and 8^x equal for all x? Use properties or values of exponential functions to justify your answer. Check by generating the curves $y = \sqrt{2^{6x}}$ and $y = 8^x$ together on your calculator or computer. Use $-0.5 \leq x \leq 1$ and a suitable range for y. Copy the graphs and put the scale on the y-axis in Figure 4.

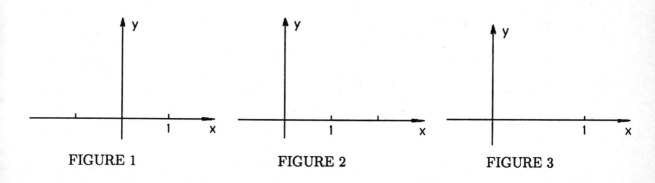

FIGURE 1   FIGURE 2   FIGURE 3

---

†This worksheet reviews properties of logarithms and exponential functions to the base b. See Calculator instructions 1B.3 for tips on solving it with a calculator.

Problem 5 Are 2^{x-x^2} and $2^x + 2^{-x^2}$ equal for all x? Use properties or values of exponential functions to justify your answer. Check by generating the curves $y = 2^{x-x^2}$ and $y = 2^x + 2^{-x^2}$ together on your calculator or computer. Use $-1 \leq x \leq 2$, and a suitable y-range. Copy the curves in Figure 5 and put the scale on the y-axis.

Problem 6 Does $\log_{10}(x^2) + \log_{10}(x^3) + \log_{10}(x^4)$ equal $9\log_{10} x$ for all x? Give your reasoning and check your answer by generating the graphs of both functions with $-1 \leq x \leq 5$ and a suitable y-range. Copy the curves in Figure 6 and put the scale on the y-axis.

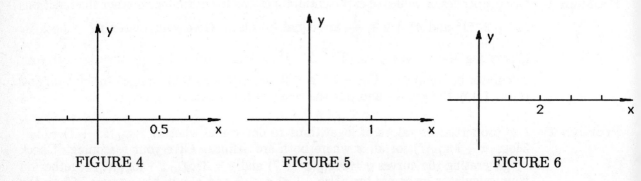

FIGURE 4   FIGURE 5   FIGURE 6

Problem 7 Does $\dfrac{6^x + 4^x}{2^x}$ equal $3^x + 2^x$ for all x? Give your reasoning and check your answer by generating the graphs of both functions with $-0.5 \leq x \leq 2$ and a suitable y-range. Copy the curves in Figure 7 and put the scale on the y-axis.

Problem 8 (a) Are $\log_{10}(x+1)$ and $\log_{10}(x^2 - 1) - \log_{10}(x - 1)$ equal for all x where both are defined? (b) Do they have the same domain? Explain. Check your answers by generating the graph of the functions separately with $-0.5 \leq x \leq 5, -0.3 \leq y \leq 1$. Copy the graph of $\log_{10}(x^2 - 1) - \log_{10}(x - 1)$ in Figure 8 and the graph of $\log_{10}(x+1)$ in Figure 9.

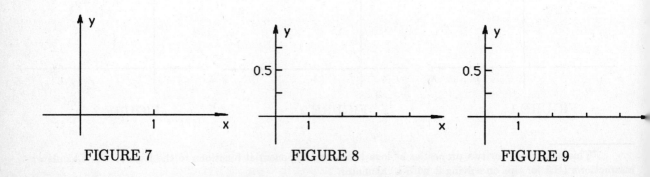

FIGURE 7   FIGURE 8   FIGURE 9

Graphing calculator workbook

Worksheet 2A.1[†]

Piecewise constant velocity

Name _____ Date _____

Others in your group _____

Instructor, Teaching Assistant, and/or Recitation section _____

Directions   Put first drafts of your calculations and answers on scratch paper. Take your time, work carefully, and discuss your solution with at least one other student before putting a final draft on this sheet or on other paper. Turn in all your work.

Problem 1   You are 25 miles north of San Diego on Interstate 5 at 1:00 AM. You drive north at one constant velocity for one hour and then at another constant velocity for two more hours. Figure 1 shows the graph of the function $s(t)$ that gives your distance from San Diego as a function of the time t, with $t = 1$ at 1:00 AM.

a.   What is your car's velocity for $1 < t < 2$?

b.   What is your car's velocity for $2 < t < 4$?

c.   The graph of the velocity v is given in Figure 2. Label the ticmarks on the vertical v-axis.

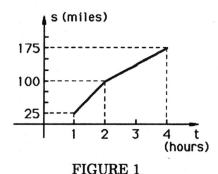

FIGURE 1   FIGURE 2

Problem 2   A couple in a sports car leaves early in the morning to drive east from El Paso, Texas. During part of the day they go 100 miles per hour but, because of car trouble, they go 50 miles per hour for another portion of the day and only 20 miles per hour during a third part of the day.

a.   Draw in Figure 3 the graph of a function $s(t)$ that gives their distance from El Paso as a function of the time t they have spent driving. Choose $s(t)$ so that they drive the 450 miles to Junction, Texas in 10 hours. There are an infinite number of possible functions.

[†]This worksheet deals with the relationship between distance and velocity in cases that do not require calculus. Velocity under more general types of motion is studied by using the derivative.

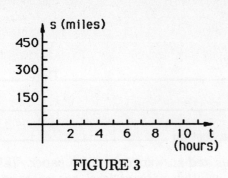
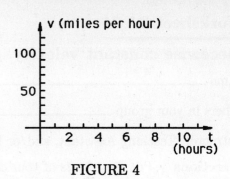

FIGURE 3   FIGURE 4

b. Draw in Figure 4 the graph of the velocity $v(t)$ corresponding to the function $s(t)$ you chose for Figure 3.

Problem 3 Figure 5 shows the graph of a ball's distance $s(t)$ from the left edge of a table as a function of the time t, measured in seconds. The ball is 10 centimeters from the left edge of the table at $t = 1$, rolls to the right for $1 < t < 6$, is stopped for $6 < t < 10$, and rolls to the left for $10 < t < 12$.

a. The ball's velocity toward the right $v(t)$ is positive for $1 < t < 6$ when it is moving toward the right, zero for $6 < t < 10$ when it is stopped, and negative for $10 < t < 12$ when it is moving toward the left. Sketch the graph of $v(t)$ in Figure 6.

b. What is the total distance the ball rolls for $1 \leq t \leq 12$?

c. What is the net distance it rolls (the distance between its positions at $t = 1$ and $t = 12$)?

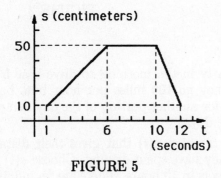
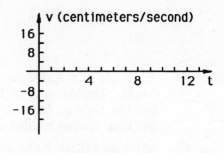

FIGURE 5   FIGURE 6

Graphing calculator workbook

Worksheet 2A.2[†]

Nonconstant velocity: A graphical approach

Name _____ Date _____

Others in your group _____

Instructor, Teaching Assistant, and/or Recitation section _____

Directions Put your initial calculations and first drafts of your conclusions on scratch paper. Take your time, work carefully, and give your conclusions in well written sentences. Discuss your solution with at least one other student before putting a final draft on this sheet or other sheets of paper, as needed. Turn in all your work.

Problem 1 A spider is crawling up a wall. Figure 1 shows the graph of its height $s(t)$ above the floor as a function of the time t measured in seconds. It is 2 feet above the floor at $t = 0$ and moves up with a varying velocity until it is 8 feet above the floor at $t = 10$.

    a. The spider's upward velocity $v(t)$ at each moment is the slope of the tangent line to the graph of $s(t)$ at that value of t. Figure 1 shows the tangent line at $t = 2$. What is the spider's upward velocity at $t = 2$? (Use the points $(2, 2.6)$ and $(6, 5.5)$ on the tangent line to find its slope.)

    b. Figure 2 shows the tangent line to the graph of $s(t)$ at $t = 7$. What is the spider's upward velocity at $t = 7$?

    c. Figure 3 shows the graph of the spider's upward velocity. Label the ticmarks on the vertical v-axis.

    d. Use the graph of the height in Figure 1 to explain why the highest point on the graph of the velocity in Figure 3 is at approximately $t = 3.5$.

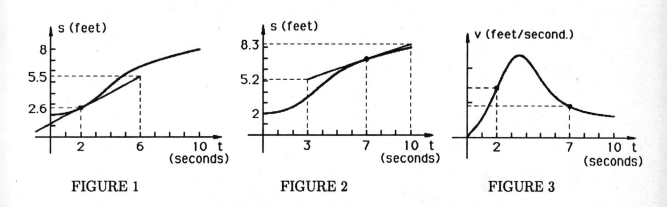

FIGURE 1   FIGURE 2   FIGURE 3

[†]This worksheet illustrates the geometric interpretation of velocity as the slope of a tangent line.

Problem 2 A model car goes east during the three seconds from $t = 0$ to $t = 3$ and then goes west back to its starting point in the three seconds from $t = 3$ to $t = 6$.

a. Figure 4 shows the graph of the car's distance $s(t)$ east of its starting point as a function of t. The slope of its tangent line at each value of t is the car's velocity toward the east at that moment. Figure 1 shows the tangent line at $t = 1$. What is the car's approximate velocity toward the east at $t = 1$?

b. Figure 5 shows the tangent line to the graph of $s(t)$ at $t = 5$. What is the car's approximate velocity toward the east at $t = 5$? It is negative because the car is moving toward the west.

c. What is the car's approximate velocity toward the west at $t = 5$?

d. Figure 6 is the graph of the car's velocity toward the east. Label the ticmarks on the vertical axis.

e. Why does the graph of the velocity cross the t-axis at $t = 3$?

f. Why is the graph of the velocity symmetric about the point $(3, 0)$ on the t-axis?

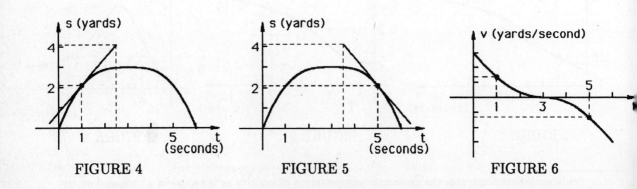

FIGURE 4   FIGURE 5   FIGURE 6

Graphing calculator workbook   71

Worksheet 2A.3†

Other rates of change: A graphical approach

Name _____ Date _____

Others in your group _____

Instructor, Teaching Assistant, and/or Recitation section _____

Directions   Put first drafts of your calculations and answers on scratch paper. Take your time, work carefully, and discuss your solution with at least one other student before putting a final draft on this sheet or on other paper. Turn in all your work.

Problem 1   A large tank contains a saline solution of 50 grams of salt per liter at time $t = 0$. Starting at that time a solution with 200 grams per liter is added. Figure 1 shows the resulting concentration of salt in the tank as a function of t.

    a.   Use the sketch to find the approximate concentration in the tank at $t = 1$:

    b.   What is the approximate concentration at $t = 5$?

    c.   As is indicated by the graph, the concentration increases as t increases and tends to 200 grams/liter as $t \to \infty$. Explain why this happens.

    d.   The rate of increase of the concentration with respect to time at each moment is the slope of the tangent line to the graph at that value of t. It is given in units of grams per liter per minute. Use the tangent line in Figure 2 to find the approximate rate of increase at $t = 1$.

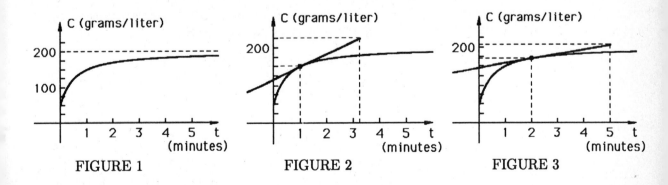

FIGURE 1   FIGURE 2   FIGURE 3

---

†This worksheet illustrates the geometric interpretation of rates of change as slopes of tangent lines.

e. Use the tangent line in Figure 3 to find the approximate rate of increase of the concentration at $t = 2$.

f. Why might you expect the rate of increase of the concentration to be less at $t = 2$ than at $t = 1$?

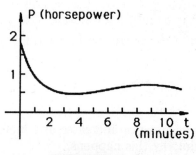

FIGURE 4

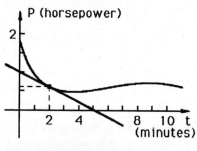

FIGURE 5

Problem 2 Figure 4 gives the amount of power P a well conditioned athlete generates cycling as a function of the number of minutes t he has cycled.

a. Approximately how long does he cycle before his power output is reduced to one horsepower?

b. When does he get his second wind?

c. The rate of increase of his power output with respect to time is measured in units of horsepower per minute. Its value at time t is the slope of the tangent line to the graph of P at time t. Figure 5 shows the tangent line at $t = 2$. What is the approximate rate of increase of his power output with respect to time at $t = 2$? What does it mean for the rate of increase to be negative?

d. Find two approximate values of t when the rate of increase of his power output is zero.

Graphing calculator workbook

Secant line program 2B

Texas Instruments TI-81

The secant line through the points at $x = A$ and $x = B$ on the graph of $f(x)$ is the line

$$(1) \qquad\qquad y = f(A) + \left[\frac{f(B) - f(A)}{B - A}\right](x - A)$$

through the points $(A, f(A))$ and $(B, f(B))$ on the graph. Instructions are given here for entering and using the following program that draws the curve and the secant line and calculates its slope, with $f(x)$ given as Y_1 in the Y= menu of the calculator. In the program, W is used for $f(A) = Y_1(A)$, V for $f(B) = Y_1(B)$, and M for the slope $M = \dfrac{V - W}{B - A}$ of the line. Consequently, the secant line has the equation $y = W + M(x - A)$.

The line numbers are given below only for reference. They are not used on the calculator.

The program (91 bytes)

1.   Prgm1: SECANT
2.   : ClrDraw
3.   : All-Off
4.   : Y_1-On
5.   : Disp "USE $Y_1(X)$"
6.   : Disp "$A =$"
7.   : Input A
8.   : Disp "$B =$"
9.   : Input B
10.   : $A \to X$
11.   : $Y_1 \to W$
12.   : $B \to X$
13.   : $Y_1 \to V$
14.   : $(V - W)/(B - A) \to M$
15.   : DrawF $W + M(X - A)$
16.   : Pause
17.   : Disp "SLOPE = "
18.   : Disp M

Entering the program

Press $\boxed{\text{2nd}}$ $\boxed{\text{QUIT}}$ to display the home screen. Then press $\boxed{\text{PRGM}}$ for the menu of programs. Press $\boxed{\blacktriangleright}$ so that EDIT is highlighted. Press $\boxed{1}$ to use program #1 or the number of an empty program if you have already used #1.

Line-by-line instructions for entering the program follow. To make a correction, move the cursor to the error and use $\boxed{\text{INS}}$ and $\boxed{\text{DEL}}$ as needed. To return to the program from other screens, press ⬛CLEAR or follow the steps in the previous paragraph. To interrupt the running of the program, press ⬛ON and then ⬛2 (QUIT).

---

1. Press ⬛S ⬛E ⬛C ⬛A ⬛N ⬛T ⬛ENTER. The calculator is automatically in alpha mode here. This title identifies the program in the program menu.

---

2. Press ⬛2nd ⬛DRAW ⬛1 ⬛ENTER. ClearDraw clears the graphics screen.

---

3. Press ⬛2nd ⬛Y-VARS ⬛◄ ⬛1 ⬛ENTER. All-Off unselects all the formulas in the $Y=$ menu so their graphs will not be generated when the program is run.

---

4. Press ⬛2nd ⬛Y-VARS ⬛► ⬛2 ⬛ENTER. This command selects Y_1 so its graph will be generated.

---

5. Press ⬛PRGM to display CTL (control) commands and then ⬛► for I/O (input/output) commands. Press ⬛1 to enter the word Disp. Press ⬛2nd ⬛A-LOCK to lock the calculator in alpha mode. Press ⬛" ⬛U ⬛S ⬛E ⬛- ⬛2nd ⬛Y-VARS ⬛1 ⬛(⬛X|T ⬛) ⬛ALPHA ⬛" ⬛ENTER with ⬛- the space on the zero key.

---

6. Press ⬛PRGM ⬛► ⬛1 for the word Disp. Press ⬛ALPHA ⬛" ⬛ALPHA ⬛A for the symbols "A and then ⬛2nd ⬛TEST to display the TEST menu. Press ⬛1 for the equal sign and ⬛ALPHA ⬛" ⬛ENTER to complete the line, which with line 7 causes the prompt $A=$? to appear when the program is run.

---

7. Press ⬛PRGM ⬛► for the I/O menu and ⬛2 ⬛ALPHA ⬛A ⬛ENTER. Here the program pauses for the user to assign a value to A.

---

8. Press ⬛PRGM ⬛► ⬛1 ⬛ALPHA ⬛" ⬛ALPHA ⬛B ⬛2nd ⬛TEST ⬛1 ⬛ALPHA ⬛" ⬛ENTER. $B=$? will appear when the program runs.

---

9. Press ⬛PRGM ⬛► ⬛2 ⬛ALPHA ⬛B ⬛ENTER. The user enters a value for B at this point in the program.

---

10. Press ⬛ALPHA ⬛A ⬛STO► ⬛X|T ⬛ENTER. This assigns the value of A to X.

Graphing calculator workbook

11. Press $\boxed{\text{2nd}}$ $\boxed{\text{Y-VARS}}$ $\boxed{1}$ $\boxed{\text{STO}\blacktriangleright}$ $\boxed{W}$ $\boxed{\text{ENTER}}$. Notice that the key $\boxed{\text{ALPHA}}$ is not needed after the $\boxed{\text{STO}\blacktriangleright}$ key. Here the function Y_1 in the Y= menu is evaluated at $X(=A)$ and the result is stored as W.

12. Press $\boxed{\text{ALPHA}}$ $\boxed{B}$ $\boxed{\text{STO}\blacktriangleright}$ $\boxed{X|T}$ $\boxed{\text{ENTER}}$ to give X the value B.

13. Press $\boxed{\text{2nd}}$ $\boxed{\text{Y-VARS}}$ $\boxed{1}$ $\boxed{\text{STO}\blacktriangleright}$ $\boxed{V}$ $\boxed{\text{ENTER}}$. $Y_1(B)$ is stored as V.

14. Use $\boxed{(}$ $\boxed{\text{ALPHA}}$ $\boxed{V}$ $\boxed{-}$ $\boxed{\text{ALPHA}}$ $\boxed{W}$ $\boxed{)}$ $\boxed{\div}$ $\boxed{(}$ $\boxed{\text{ALPHA}}$ $\boxed{B}$ $\boxed{-}$ $\boxed{\text{ALPHA}}$ $\boxed{A}$ $\boxed{)}$ $\boxed{\text{STO}}$ $\boxed{M}$ $\boxed{\text{ENTER}}$. The slope $(V - W)/(B - A)$ of the secant line through $(A, W) = (A, Y_1(A))$ and $(B, V) = (B, Y_1(B))$ on the graph is calculated and stored as M.

15. Press $\boxed{\text{2nd}}$ $\boxed{\text{DRAW}}$ $\boxed{6}$ $\boxed{\text{ALPHA}}$ $\boxed{W}$ $\boxed{+}$ $\boxed{\text{ALPHA}}$ $\boxed{M}$ $\boxed{(}$ $\boxed{X|T}$ $\boxed{-}$ $\boxed{\text{ALPHA}}$ $\boxed{A}$ $\boxed{)}$ $\boxed{\text{ENTER}}$. The secant line $y = W + M(x - A)$ is drawn.

16. Press $\boxed{\text{PRGM}}$ $\boxed{6}$ $\boxed{\text{ENTER}}$ to insert a PAUSE command. This stops the program until the user presses $\boxed{\text{ENTER}}$ to continue.

17. Press $\boxed{\text{PRGM}}$ $\boxed{\blacktriangleright}$ $\boxed{1}$ and then $\boxed{\text{2nd}}$ $\boxed{\text{A-LOCK}}$ to lock the calculator in alpha mode. Press $\boxed{``}$ $\boxed{S}$ $\boxed{L}$ $\boxed{O}$ $\boxed{P}$ $\boxed{E}$ $\boxed{\text{2nd}}$ $\boxed{\text{TEST}}$ $\boxed{1}$ $\boxed{\text{ALPHA}}$ $\boxed{``}$ $\boxed{\text{ENTER}}$. SLOPE = appears when the program is run.

18. Press $\boxed{\text{PRGM}}$ $\boxed{\blacktriangleright}$ $\boxed{1}$ $\boxed{\text{ALPHA}}$ $\boxed{M}$ $\boxed{\text{ENTER}}$ for the last command, which displays the slope of the line on the home screen.

Using the program

Before running the program, select the ranges of x and y for the graph and put the formula for the function to be studied as Y_1 in the Y= menu.

Example   Run the program for the six secant lines to the parabola $y = 2x - \frac{1}{4}x^2$ through the points at $x = 2$ and **(a)** $x = 5$; **(b)** $x = 4$; **(c)** $x = 3$; **(d)** $x = 2.1$; **(e)** $x = 2.001$; and **(f)** $x = 2.0001$. Use $-0.5 \le x \le 8.5$ and $-1 \le y \le 5$.

Solution   Press $\boxed{\text{RANGE}}$ and set x-min $= -0.5$, x-max $= 8.5$, x-scale $= 1$, y-min $= -1$, y-max $= 5$ and y-scale $= 1$.

Press $\boxed{\text{Y} =}$ $\boxed{2}$ $\boxed{X|T}$ $\boxed{-}$ $\boxed{X|T}$ $\boxed{x^2}$ $\boxed{\div}$ $\boxed{4}$ and delete any extra symbols on the line to have $Y_1 = 2X - X^2/4$.

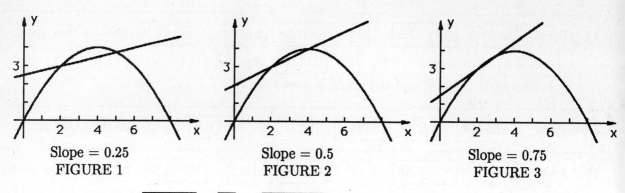

Slope = 0.25
FIGURE 1

Slope = 0.5
FIGURE 2

Slope = 0.75
FIGURE 3

(a) Press [PRGM] [1] [ENTER] to run the program if it is program #1. A reminder that the function you want to study should be entered as $Y_1(X)$ should appear, followed by the prompt $A =?$ Press [2] [ENTER] to assign the value 2 to A and then [5] [ENTER] to assign the value 5 to B. The curve and the secant line should appear as in Figure 1. Press [ENTER] to obtain the value 0.25 for the slope of the secant line.

If there is an error in the program that the calculator can locate, an error message will be given with the number of the type of error as described in the owner's manual. Press [1] to go to the line of the program with the error so you can correct it. If the program does not do what you expect but no error message is displayed, compare the program in the calculator with the listing above.

(b) Press [ENTER] or [PRGM] [1] [ENTER] to run the program again. Use $A = 2$ and $B = 4$ to obtain the drawing in Figure 2 and the slope 0.5.

(c) Use $A = 2$ and $B = 3$ for the drawing in Figure 3 and the slope 0.75.

(d) Set $A = 2$ and $B = 2.1$ for the drawing in Figure 4 and the slope 0.975.

(e) Set $A = 2$ and $B = 2.001$ for the drawing in Figure 5 and the slope 0.99975.

(f) Finally with $A = 2$ and $B = 2.00001$ you obtain the drawing in Figure 6 and 0.9999975 for the slope of the secant line. The secant lines in the last two figures are indistinguishable because their slopes are very close to each other.

The tangent line at $x = 2$ is the limiting position of the secant line through the points at $x = 2$ and B as B approaches 2. Its slope is 1.

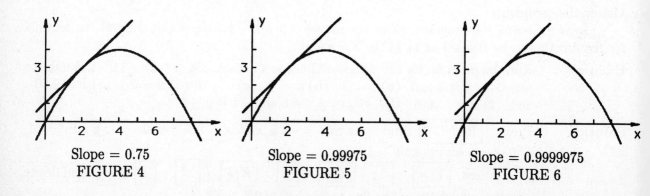

Slope = 0.75
FIGURE 4

Slope = 0.99975
FIGURE 5

Slope = 0.9999975
FIGURE 6

Graphing calculator workbook

Worksheet 2C.1[†]

Tangent lines to y = x^n: Numerical experiments

Name _____ Date _____

Others in your group _____

Instructor, Teaching Assistant, and/or Recitation section _____

Directions   Put first drafts of your calculations and answers on scratch paper. Take your time, work carefully, and discuss your solution with at least one other student before putting a final draft on this sheet or on other paper. Turn in all your work.

Problem 1a   Use Secant line program 2B to generate the curve $y = x^2$ and secant lines to it through **(i)** $x = 1$ and $x = 2$, **(ii)** $x = 1$ and $x = 1.1$, **(iii)** $x = 1$ and $x = 1.001$, and **(iv)** $x = 1$ and $x = 1.00001$. Use $-1 \le x \le 3.5, -0.5 \le y \le 2.5$ for the graphs. The slopes of the secant lines should be the numbers in the first row of the table below. Copy the curve and the secant line for $x = 1.00001$ in Figure 1.

   b.   Pick an integer $n > 2$. Follow the instructions for Problem 1 with $y = x^n$ in place of $y = x^2$. Put the value of n and the slopes in the second row of the table. Copy the curve and the secant line for $x = 1.00001$ in Figure 2.

SLOPE OF THE SECANT LINE TO $y = x^n$ THROUGH $x = 1 (= A)$ AND $x = B$

n	$A=1, B=2$	$A=1, B=1.1$	$A=1, B=1.001$	$A=1, B=1.00001$
2	3	2.1	2.001	2.00001

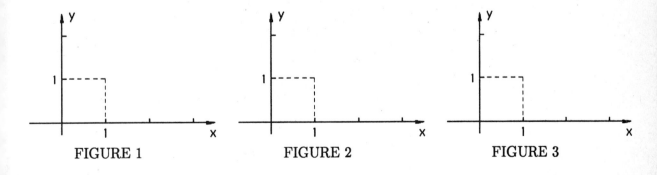

FIGURE 1   FIGURE 2   FIGURE 3

[†]This worksheet uses the definition of the derivative and Secant line program 2B to predict the formula for the derivative of x^n at $x = 1$. This formula is established in your textbook through algebraic manipulations of expressions for the slopes of secant lines.

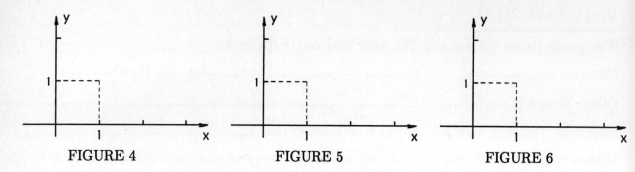

FIGURE 4　　　　　　　　FIGURE 5　　　　　　　　FIGURE 6

c. Pick a fraction $n > 1$ (not an integer). Follow the instructions for Problem 1 with $y = x^n$ in place of $y = x^2$. Put n and the slopes in the third row of the table and copy the curve and the secant line for $x = 1.00001$ in Figure 3.

d. Pick a positive fraction $n < 1$. Follow the instructions for Problem 1 with $y = x^n$ in place of $y = x^2$. Put n and the slopes in the fourth row of the table and copy the curve and the secant line for $x = 1.00001$ in Figure 4.

e. Pick a negative fraction $n > -1$. Follow the instructions for Problem 1 with $y = x^n$ in place of $y = x^2$. Put n and the slopes in the fifth row of the table and copy the curve and the secant line for $x = 1.00001$ in Figure 5.

f. Pick a negative integer $n < -1$. Follow the instructions for Problem 1 with $y = x^n$ in place of $y = x^2$. Put n and the slopes in the sixth row of the table and copy the curve and the secant line for $x = 1.00001$ in Figure 2.

Problem 2　The tangent line to $y = x^n$ at $x = 1$ is the limiting position of the secant line through $x = 1$ and $x = B$ as $B \to 1$, and the slope of the tangent line is the limit of the slope of the secant line. Use the table to predict the slopes of the tangent lines to $y = x^n$ at $x = 1$ for the values of n you used in Problem 1.

Problem 3　There is a simple formula for the slope of the tangent line to $y = x^n$ at $x = 1$. Based on the results of Problem 2, what do you think the formula is?

Worksheet 2C.2†

Tangent lines to $y = \cos x$ and $y = \sin x$: Numerical experiments

Name _____ Date _____

Others in your group _____

Instructor, Teaching Assistant, and/or Recitation section _____

Directions Put first drafts of your calculations and answers on scratch paper. Take your time, work carefully, and discuss your solution with at least one other student before putting a final draft on this sheet or on other paper. Turn in all your work.

Problem 1a Generate the graph of $y = \sin x$ on your calculator or computer for $-0.5 \leq x \leq 7$, $-1.5 \leq y \leq 1.5$ and copy it in Figure 1.

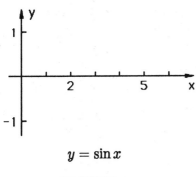

$y = \sin x$

FIGURE 1

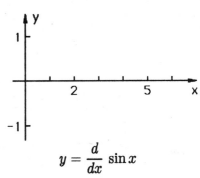

$y = \dfrac{d}{dx} \sin x$

FIGURE 2

b. Use Secant line program 2B with the x- and y-ranges from part (a) to complete Table 1 with the slopes, accurate to two-decimal places, of the secant lines to $y = \sin x$ through the points at $x = A$ and $x = B = A + 0.001$ for $A = 0, 0.5, 1, 1.5, 2, 2.5, 3, 3.5, 4, 4.5, 5, 5.5, 6,$ and 6.5.

A	B	$\dfrac{\sin(B) - \sin(A)}{B - A}$	A	B	$\dfrac{\sin(B) - \sin(A)}{B - A}$
0.0	0.001	1.00	3.5	3.501	−0.94
0.5	0.501		4.0	4.001	
1.0	1.001		4.5	4.501	
1.5	1.501		5.0	5.001	
2.0	2.001		5.5	5.501	
2.5	2.501		6.0	6.001	
3.0	3.001		6.5	6.501	

TABLE 1

---

†This worksheet uses the definition of the derivative and Secant line program 2B to predict formulas for the derivatives of $\sin x$ and $\cos x$. These formulas are established in your textbook by first proving that $\sin x / x \to 1$ as $x \to 0$ and then using trigonometric identities or geometry. See Calculator instructions 1B.3 for tips on solving these problems with a calculator.

c. Each slope $\dfrac{\sin(B) - \sin(A)}{B - A}$ of a secant line in Table 1 is close to the derivative $\dfrac{d}{dx}\sin x$ at $x = A$ since B is close to A. Plot these 14 approximate values of the derivative in Figure 2.

Problem 2a Generate the graph of $y = \cos x$ on your calculator or computer for $-0.5 \leq x \leq 7$, $-1.5 \leq y \leq 1.5$ and copy it in Figure 3.

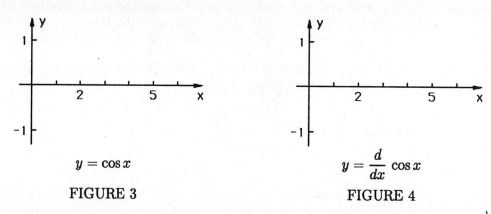

$y = \cos x$

FIGURE 3

$y = \dfrac{d}{dx}\cos x$

FIGURE 4

b. Use Secant line program 2B with the x- and y-ranges from part (a) to complete Table 2 with the slopes, accurate to two decimal places, of the secant lines to $y = \cos x$ through the points at $x = A$ and $x = B = A + 0.001$ for $A = 0, 0.5, 1, 1.5, 2, 2.5, 3, 3.5, 4, 4.5, 5, 5.5, 6$, and 6.5.

c. Each slope $\dfrac{\cos(B) - \cos(A)}{B - A}$ of a secant line in Table 2 is close to the derivative $\dfrac{d}{dx}\cos x$ at $x = A$. Plot these 14 approximate values of the derivative in Figure 4.

Problem 3 There are simple formulas for the derivatives of $\sin x$ and $\cos x$. See if you can guess what they are by comparing Figures 1 through 4.

A	B	$\dfrac{\cos(B) - \cos(A)}{B - A}$	A	B	$\dfrac{\cos(B) - \cos(A)}{B - A}$
0.0	0.001	0.00	3.5	3.501	0.35
0.5	0.501		4.0	4.001	
1.0	1.001		4.5	4.501	
1.5	1.501		5.0	5.001	
2.0	2.001		5.5	5.501	
2.5	2.501		6.0	6.001	
3.0	3.001		6.5	6.501	

TABLE 2

Graphing calculator workbook

Worksheet 2D.1[†]

Calculus with functions given approximately by graphs

Name _____ Date _____

Others in your group _____

Instructor, Teaching Assistant, and/or Recitation section _____

Directions   Put first drafts of your calculations and answers on scratch paper. Take your time, work carefully, and discuss your solution with at least one other student before putting a final draft on this sheet or on other paper. Turn in all your work.

Problem 1   Figure 1 shows the graph of a function $f(x)$. Draw approximate tangent lines at $x = 1$ and $x = 3$ and estimate their slopes to find approximate values of $\dfrac{df}{dx}(1)$ and $\dfrac{df}{dx}(3)$.

Problem 2   Find the approximate values of x where the function $g(x)$ of Figure 2 has a zero derivative. (Find where its tangent line is horizontal.) Then find the approximate value of x where the derivative is 1. (Draw a line of slope 1 on the page near the graph and determine where the tangent line is parallel to that line.)

Problem 3   What is the approximate maximum value of $\dfrac{dh}{dx}(x)$ for $0 \leq x \leq 4$ where $h(x)$ is the function of Figure 3? Notice that the scales on the x- and y-axes are different.

Problem 4a   What is the approximate value at $x = 2$ of the function $U(x)$ whose graph is in Figure 4? What is the approximate value of its derivative at $x = 2$?

   b.   Use the results from part (a) and the product rule to find the approximate value of the derivative of $x^3 U(x)$ at $x = 2$.

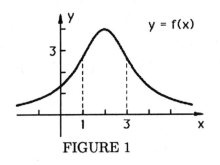

FIGURE 1

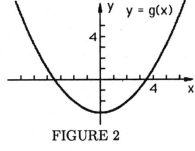

FIGURE 2

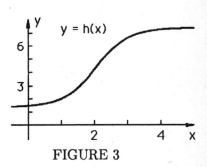

FIGURE 3

[†]These problems use the rules for differentiating powers of x, linear combinations, products, and quotients with approximate derivatives obtained by estimating slopes of tangent lines, as in Worksheets 2A.2 and 2A.3.

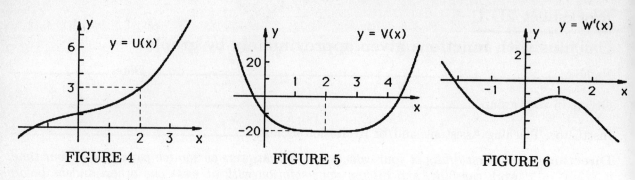

FIGURE 4   FIGURE 5   FIGURE 6

Problem 5a   What is the approximate derivative at $x = 2$ of the function $V(x)$ of Figure 5?

   b.   Use the graph and the result of part (a) to find the approximate derivative of $\dfrac{V(x)}{x}$ at $x = 2$.

Problem 6   Figure 6 shows the graph of $\dfrac{dW}{dx}(x)$. Sketch the graph of $\dfrac{d}{dx}[W(x) + 3x]$ in the same drawing.

Problem 7   Figures 7 through 9 show the graphs of $P(x)$, $Q(x)$, and $R(x)$. Which of Figures 10 through 12 is the graph of $\dfrac{dP}{dx}$? Of $\dfrac{dQ}{dx}$? Of $\dfrac{dR}{dx}$? Explain.

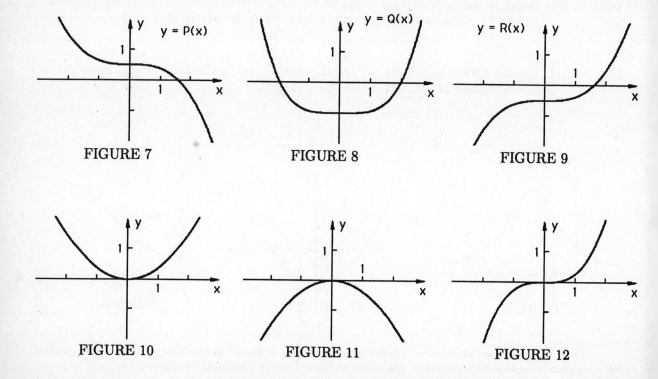

FIGURE 7   FIGURE 8   FIGURE 9

FIGURE 10   FIGURE 11   FIGURE 12

Worksheet 2D.2†

Derivatives of composite functions: Initial investigations

Name _____ Date _____

Others in your group _____

Instructor, Teaching Assistant, and/or Recitation section _____

Directions Put first drafts of your calculations and answers on scratch paper. Take your time, work carefully, and discuss your solution with at least one other student before putting a final draft on this sheet or on other paper. Turn in all your work.

Problem 1 Figure 1 shows the graph of the total sales $M(t)$ of Better Mouse Traps as a function of time t over a period of four months. The traps are sold at constant rates for $0 < t < 2$, for $2 < t < 3$ and for $3 < t < 4$. Because the traps do not work very well, one is returned for every one that is sold in the third month and more are returned in the fourth month than are sold. The rate of change $\dfrac{dM}{dt}$ of the number of mouse traps sold with respect to time is given in units of mouse traps per year.

a. What is $\dfrac{dM}{dt}$ for $0 < t < 2$?

b. What is $\dfrac{dM}{dt}$ for $2 < t < 3$?

c. What is $\dfrac{dM}{dt}$ for $3 < t < 4$?

Problem 2 The designer earns $2.00 per mouse trap for the first 1000 and $0.50 per mouse trap for any over 1000 that are sold. Figure 2 shows the graph of his income $I(M)$ as a function of the number sold. The rate of change $\dfrac{dI}{dM}$ of his income with respect to the number of mousetraps sold is measured in units of dollars per mousetrap.

a. What is $\dfrac{dI}{dM}$ for $0 < M < 1000$?

b. What is $\dfrac{dI}{dM}$ for $M > 1000$?

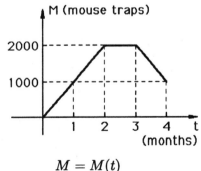

$M = M(t)$

FIGURE 1

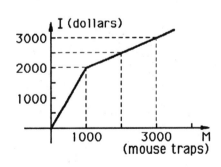

$I = I(M)$

FIGURE 2

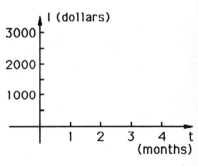

$I = I(M(t))$

FIGURE 3

---

†This short worksheet leads to the "chain rule" in an example where the derivatives can be found as slopes of the line segments that form the graphs. The general chain rule is derived in your textbook.

84   Graphing calculator workbook

Problem 3   The designer's income is given as a function of time by the composite function $I(M(t))$. Find its values at $t = 0, 1, 2, 3$, and 4:

    a. $I(M(0)) = I(\underline{\hspace{1cm}}) =$

    b. $I(M(1)) = I(\underline{\hspace{1cm}}) =$

    c. $I(M(2)) = I(\underline{\hspace{1cm}}) =$

    d. $I(M(3)) = I(\underline{\hspace{1cm}}) =$

    e. $I(M(4)) = I(\underline{\hspace{1cm}}) =$

Problem 4   Sketch the graph of I as a function of t in Figure 3.

Problem 5   The rate of change $\dfrac{dI}{dt}$ of the designer's income with respect to time is measured in units of dollars per year.

    a.   What is $\dfrac{dI}{dt}$ for $0 < t < 1$?

    b.   What is $\dfrac{dI}{dt}$ for $1 < t < 2$?

    c.   What is $\dfrac{dI}{dt}$ for $2 < t < 3$?

    d.   What is $\dfrac{dI}{dt}$ for $3 < t < 4$?

Problem 5   How are the values of $\dfrac{dI}{dt}$ in part (e) related to the values of $\dfrac{dI}{dM}$ and $\dfrac{dM}{dt}$ in parts (a) and (b)? Express the relationship as an equation involving $\dfrac{dI}{dt}$, $\dfrac{dI}{dM}$ and $\dfrac{dM}{dt}$. This is an example of what is called the "chain rule."

Graphing calculator workbook

Worksheet 2D.3†

Derivatives of composite functions: Examples with powers

Name _____ Date _____

Others in your group _____

Instructor, Teaching Assistant, and/or Recitation section _____

Directions *Put first drafts of your calculations and answers on scratch paper. Take your time, work carefully, and discuss your solution with at least one other student before putting a final draft on this sheet or on other paper. Turn in all your work.*

Problem 1a The graph of $y = \sqrt{x}$ is shown in Figure 1 with its tangent line at $x = 4$. Use a differentiation formula to find the derivative $\left.\dfrac{dy}{dx}\right|_{x=4}$ and put its value under the drawing. Check your result by estimating the slope of the line in the drawing.

b. The graph of $z = \frac{1}{4}y^3$ is shown in Figure 2 with its tangent line at $y = 2$. Use a differentiation formula to find the derivative $\left.\dfrac{dz}{dy}\right|_{y=2}$ and put the value under the drawing. Check by estimating the slope of the tangent line in the drawing.

c. Substitute the formula for y in terms of x from part (a) into the formula for z in terms of y from part (b) to express z as a function of x. Check your answer by generating the graph of z as a function of x for $-1 \leq x \leq 8$, $-1 \leq z \leq 5$ and comparing the result with the curve in Figure 3.

d. Use differentiation formulas and the equation from part (c) to find the derivative $\left.\dfrac{dz}{dx}\right|_{x=4}$ and put the value under Figure 3. Check by estimating the slope of the tangent line.

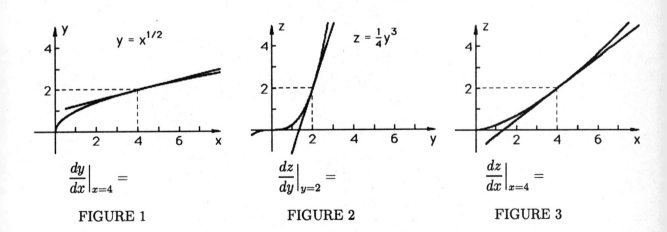

FIGURE 1 $\qquad$ FIGURE 2 $\qquad$ FIGURE 3

---

†This worksheet illustrates the geometric interpretation of the chain rule as a relation between slopes of tangent lines and the chain rule for a doubly composite function $w(z(y(x)))$. It uses the chain rule and the formula for differentiating Ax^n for constants A and n.

e. The chain rule for differentiating $z(y(x))$ with respect to x reads $\dfrac{dz}{dx} = \dfrac{dz}{dy}\dfrac{dy}{dx}$ if the values of the variables are omitted. Use this formula to explain how the slopes of the tangent lines in Figures 1 through 3 are related.

Problem 2a Figure 4 shows the graph of $w = 4/z$ and its tangent line at $z = 2$. Use a differentiation formula to find the derivative $\left.\dfrac{dw}{dz}\right|_{z=2}$ and put its value under the drawing. Check your answer by estimating the slope of the tangent line.

b. Substitute the formula for z in terms of x from Problem 1c into the formula for w in terms of z from part (a) of this problem to express w as a function of x. Check your formula by generating the graph for $-1 \leq x \leq 8$, $-1 \leq w \leq 6$ and comparing the result with the curve in Figure 5.

c. Use differentiation formulas and the equation from part (b) to find the derivative $\left.\dfrac{dw}{dx}\right|_{x=4}$ and put its value under the drawing.

Problem 3 In Problem 1 we used the rule $\dfrac{dz}{dx} = \dfrac{dz}{dy}\dfrac{dy}{dx}$. Give the analogous formula for expressing $\dfrac{dw}{dx}$ in terms of $\dfrac{dw}{dz}, \dfrac{dz}{dy}$, and $\dfrac{dy}{dz}$. This type of formula is what gives the "chain" rule its name.

Problem 4 Use the chain rule from Problem 3 to explain how the slopes

$$\left.\dfrac{dy}{dx}\right|_{x=4}, \quad \left.\dfrac{dz}{dy}\right|_{y=2}, \quad \left.\dfrac{dw}{dz}\right|_{z=2}, \quad \text{and} \quad \left.\dfrac{dw}{dx}\right|_{x=4}$$

of the tangent lines in Figures 1, 2, 4, and 5 are related.

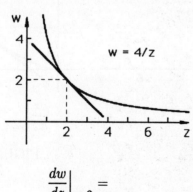

$\left.\dfrac{dw}{dz}\right|_{z=2} =$

FIGURE 4

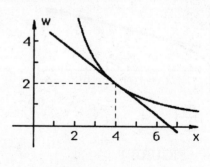

$\left.\dfrac{dw}{dx}\right|_{x=4} =$

FIGURE 5

Worksheet 2D.4†

A closer look at some derivatives.

Name _____ Date _____

Others in your group _____

Instructor, Teaching Assistant, and/or Recitation section _____

Directions Put your initial calculations and first drafts of your conclusions on scratch paper. Take your time, work carefully, and give your conclusions in well written sentences. Discuss your solution with at least one other student before putting a final draft on this sheet or other sheets of paper, as needed. Turn in all your work.

Problem 1 Generate the graphs of $\dfrac{(x+0.2)^3 - x^3}{0.2}$ and $3x^2$ together on your calculator or computer and copy them in Figure 1. Use $-2.5 \le x \le 2.5$, $-3 \le y \le 15$, and y-scale $= 5$. Then, generate the graphs of $\dfrac{\sin(x+0.3) - \sin x}{0.3}$ and $\cos x$ for $-4 \le x \le 4$, $-1.5 \le y \le 1.5$ and copy them in Figure 2. Explain the results.

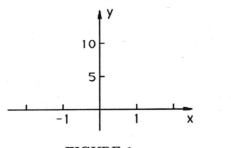

FIGURE 1   FIGURE 2

Problem 2 The function $A(x)$ of Figure 3 does not have a derivative at $x = 1$ because its graph has a vertical tangent line there. What is $\displaystyle\lim_{x \to 1} \dfrac{A(x) - A(1)}{x - 1}$?

Problem 3 The function $B(x)$ of Figure 4 has a corner at $x = 1$ because the one-sided limits $\displaystyle\lim_{x \to 1^-} \dfrac{B(x) - B(1)}{x - 1}$ and $\displaystyle\lim_{x \to 1^+} \dfrac{B(x) - B(1)}{x - 1}$ exist but are different. What are their approximate values? $B(x)$ does not have a derivative and its graph does not have a tangent line at $x = 1$. Explain.

Problem 4 The function $C(x)$ of Figure 5 is discontinuous at $x = 1$. What is $\displaystyle\lim_{x \to 1^-} \dfrac{C(x) - C(1)}{x - 1}$ and what is the approximate value of $\displaystyle\lim_{x \to 1^+} \dfrac{C(x) - C(1)}{x - 1}$? $C(x)$ does not have a derivative and its graph does not have a tangent line at $x = 1$. Explain.

---

†These problems explore the definition of derivative and the principle that a tangent line is the line that best approximates a graph near the point of tangency.

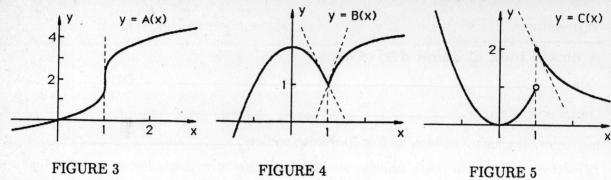

FIGURE 3 FIGURE 4 FIGURE 5

Problem 5 The curve $y = 1+(x-1)-(x-1)^3$ has a tangent line at $x = 1$ and $y = 1-2|x-1|$ does not. Generate these curves together on your calculator or computer **(a)** for $-1 \le x \le 3, -1 \le y \le 2$, **(b)** for $0.4 \le x \le 1.6, 0.6 \le y \le 1.4$, and **(c)** for $0.97 \le x \le 1.03, 0.98 \le y \le 1.02$. Copy them in Figures 6 through 8, label them, and explain the results.

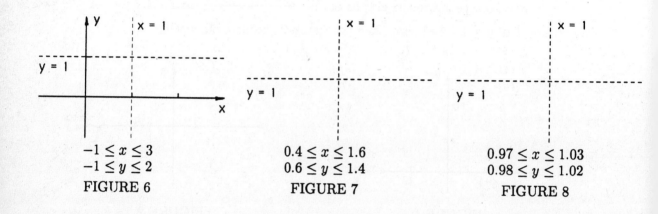

$-1 \le x \le 3$
$-1 \le y \le 2$
FIGURE 6

$0.4 \le x \le 1.6$
$0.6 \le y \le 1.4$
FIGURE 7

$0.97 \le x \le 1.03$
$0.98 \le y \le 1.02$
FIGURE 8

Problem 6 The tangent line to the parabola $y = 1 + (x-1)^2$ at $x = 1$ is $y = 1$. It is not $y = 1 + 0.01(x-1)$. Generate the parabola and the two lines together **(a)** for $-0.5 \le x \le 2, -0.5 \le y \le 3$, **(b)** for $0.99 \le x \le 1.01, 0.99 \le y \le 1.01$, and **(c)** for $0.99 \le x \le 1.01, 0.9999 \le y \le 1.0001$. Copy the graphs in Figures 9 through 11 and explain the results.

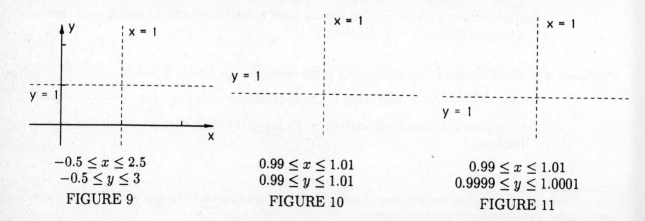

$-0.5 \le x \le 2.5$
$-0.5 \le y \le 3$
FIGURE 9

$0.99 \le x \le 1.01$
$0.99 \le y \le 1.01$
FIGURE 10

$0.99 \le x \le 1.01$
$0.9999 \le y \le 1.0001$
FIGURE 11

Graphing calculator workbook

Calculator instructions 2E: Graphs of approximate derivatives

Texas Instruments TI-81

If $f(x)$ has a derivative at x, then the derivative $f'(x)$ is the limit of the difference quotient

(1)
$$\frac{f(x + \Delta x) - f(x)}{\Delta x}$$

as Δx tends to 0. It is also the limit of the "centered" difference quotient

(2)
$$\frac{f(x + \Delta x) - f(x - \Delta x)}{2 \Delta x}.$$

Either of these difference quotients is an approximation of the derivative $f'(x)$ if Δx is small. Because of this, they are called "approximate" or "numerical" derivatives.

The centered difference quotient (2) is generally more accurate than the usual difference quotient (1) for the same value of Δx, because the secant line through $(x - \Delta x, f(x - \Delta x), x + \Delta x, f(x + \Delta x))$ has slope that is closer to that of the tangent line. (Draw tangent lines and both types of secant lines to a few graphs to see why this is the case.)

The calculator has a built-in procedure for generating graphs of approximate derivatives calculated as centered difference quotients. The next example shows how to use it.

Example 1   Use the NDeriv procedure with $\Delta x = 0.8$ to draw the graph of the approximate derivative of x^3 for $-1.5 \leq x \leq 1.5, -3 \leq y \leq 8$. Then generate the graph of the approximate derivative and the exact derivative $3x^2$ on the same screen.

Solution   Press ⎹Y=⎸ ⎹X|T⎸ ⎹∧⎸ ⎹3⎸ and delete any extra symbols to have the first line of the Y= menu read $Y_1 = X \wedge 3$. Move the cursor to the equal sign on the first line and press ⎹ENTER⎸ to unselect Y_1. Press ⎹▼⎸ ⎹▶⎸ to move to the next line. Press ⎹MATH⎸ ⎹8⎸ to put NDeriv(after $Y_2 =$. Press ⎹2nd⎸ ⎹Y-vars⎸ ⎹1⎸ ⎹ALPHA⎸ ⎹,⎸ ⎹0⎸ ⎹.⎸ ⎹8⎸ ⎹)⎸ and delete any extra symbols to have line 2 read $Y_2 = \text{NDeriv}(Y_1, 0.8)$ with its equal sign highlighted. This defines Y_2 to be the numerical derivative (2) with Y_1 for f and with $\Delta x = 0.8$. The equal signs for Y_1, Y_3, and Y_4 should not be highlighted.

Press ⎹RANGE⎸ and set x-min $= -1.5$, x-max $= 1.5$, x-scale $= 1$, y-min $= -3$, y-max $= 8$, and y-scale $= 1$ in the Range menu. Press ⎹GRAPH⎸ to generate the graph of the approximate derivative of x^3,

$$\frac{(x + 0.8)^3 - (x - 0.8)^3}{2(0.8)}.$$

Press $\boxed{Y=}$, move the cursor to line 3, press $\boxed{3}$ $\boxed{X|T}$ $\boxed{x^2}$ and delete any extra symbols to have line 3 read $Y_3 = 3X^2$. The equal signs on lines 2 and 3 should be highlighted. Press $\boxed{MODE}$, and select **Sequence** if it is not already selected. Press $\boxed{GRAPH}$ to generate first the graph of the approximate derivative and then the graph of the exact derivative of x^3, which is $3x^2$.

A fairly large value of Δx was used in this example so the graphs of the approximate and exact derivatives would be distinguishable. See what happens when the graphs of the exact and approximate derivatives are drawn with $\Delta x = 0.01$ instead of $\Delta x = 0.8$ by setting $Y_2 = \text{NDeriv}(Y_1, 0.01)$.

Generating approximate graphs of second derivatives

To sketch the approximate graph of the second derivative of $f(x)$, put $Y_1 = f(X)$, $Y_2 = \text{NDeriv}(Y_1, \Delta x)$, and $Y_3 = \text{NDeriv}(Y_2, \Delta x)$ with a small Δx, and select only the graph of Y_3 to be plotted.

Finding approximate values of derivatives

To find the approximate decimal value of the derivative of Y_1 at x, put an expression for x in the home screen and press $\boxed{STO\blacktriangleright}$ $\boxed{X|T}$ $\boxed{ENTER}$ and then $\boxed{2nd}$ $\boxed{Y\text{-VARS}}$ $\boxed{2}$ $\boxed{ENTER}$.

Checking formulas for derivatives

You can also use NDeriv to verify formulas you find for derivatives. Set Y_1 to be the function you have differentiated, set $Y_2 = \text{NDeriv}(Y_1, \Delta x)$ with a fairly large Δx such as 0.5, and define Y_3 by your formula for the derivative. Generate Y_2 and Y_3 together. They should be almost the same, but distinguishable because of the large Δx.

Graphing calculator workbook

Worksheet 2F.1[†]

Forming polynomials from monomials

Name _____ Date _____

Others in your group _____

Instructor, Teaching Assistant, and/or Recitation section _____

Directions Put your initial calculations and first drafts of your conclusions on scratch paper. Take your time, work carefully, and give your conclusions in well written sentences. Discuss your solution with at least one other student before putting a final draft on this sheet or other sheets of paper, as needed. Turn in all your work.

Problem 1 Use x-min $= -4$, x-max $= 4$, y-min $= -4$, and y-max $= 10$ in this problem.

a. Use a pencil to draw the lines $y = 3$ and $y = 3 - x$ in Figure 1 without using a calculator or computer. Notice the unequal scales on the axes. Next, generate the lines with your calculator or computer and correct your drawing if necessary. How does changing $y = 3$ to $y = 3 - x$ change the line? Use the formulas to explain why this occurs.

b. Use a pencil to draw the line $y = 3 - x$ and the parabola $y = 3 - x + x^2$ in Figure 2 without using a calculator or computer. Plot the points at $x = -2, 0, 2$. Then generate the parabola on your calculator or computer and correct your drawing if necessary. How does adding x^2 change the shape of the graph? Use the formulas to explain why this occurs.

c. Generate the curve $y = 3 - x + x^2 + x^3$ with the parabola $y = 3 - x + x^2$ on your calculator or computer and copy them in Figure 3. How does adding x^3 change the shape of the graph? Explain.

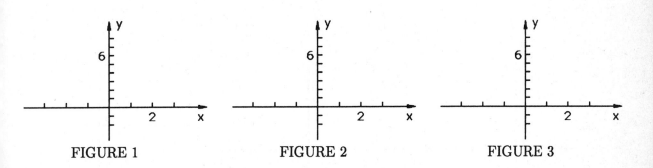

FIGURE 1   FIGURE 2   FIGURE 3

[†]These problems do not use any calculus. They illustrate the principles that the lower order terms of a polynomial determine the shape of its graph near $x = 0$ and its highest order term determines its limits as $x \to \pm\infty$.

Problem 2 Use $-3 \leq x \leq 3, -2 \leq y \leq 5$ in this problem.

  a. Generate the line $y = 1$ and the parabola $y = 1 + x^2$ together on your calculator or computer and copy them in Figure 4. Then generate $y = 1 + x^2$ and $y = 1 + x^2 - \frac{1}{5}x^4$ and draw them in Figure 5.

  b. How does adding x^2 change the graph? How does subtracting $\frac{1}{5}x^4$ change it? Use the formulas to explain why these changes occur.

Problem 3 Experiment with various positive values of a and b and various choices of ranges of x and y until you obtain a good looking graph of $y = ax^3 - bx^5$ on your calculator or computer. Generate $y = ax^3 - bx^5$ and $y = ax^3$ together, copy the two curves in Figure 6, and put the scale on the y-axis. Then explain how and why subtracting the term bx^5 changes the graph.

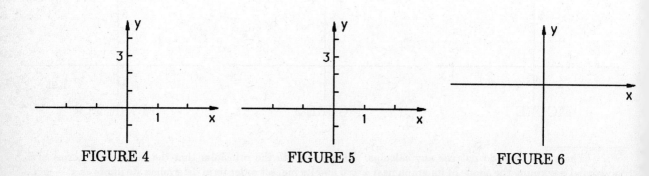

FIGURE 4   FIGURE 5   FIGURE 6

Worksheet 2F.2†

Factored polynomials and rational functions

Name _____ Date _____

Others in your group _____

Instructor, Teaching Assistant, and/or Recitation section _____

Directions Put your initial calculations and first drafts of your conclusions on scratch paper. Take your time, work carefully, and give your conclusions in well written sentences. Discuss your solution with at least one other student before putting a final draft on this sheet or other sheets of paper, as needed. Turn in all your work.

Problem 1 Figure 1 shows the graph of

$$y = (x-a)^m(x-b)^n$$

for certain integers a, b, m, and n.

a. Use the sketch to find a, b, m, and n. Check by generating the curve on your calculator or computer with x-min $= -3$, x-max $= 4$, x-scale $= 1$, y-min $= -15$, y-max $= 15$, and y-scale $= 5$.

b. Explain how these values of a, b, m, and n determine the shape of the curve.

Problem 2 Figure 2 shows the curve

$$y = (x+1)^p(x-1)^q(x-2)^r$$

for certain integers p, q, and r.

a. Use the sketch to find p, q, and r. Check by generating the curve on your calculator or computer. Use x-min $= -2$, x-max $= 3$, x-scale $= 1$, y-min $= -5$, y-max $= 5$, and y-scale $= 1$.

b. How do these values of p, q, and r determine the shape of the curve?

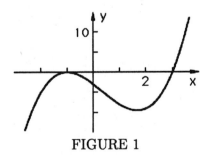
FIGURE 1

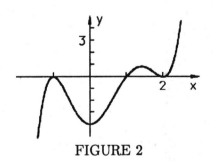
FIGURE 2

---

†These problems do not use any calculus. They illustrate how the factors in a factored polynomial or rational function affect the shape of its graph.

Problem 3   Figure 3 shows the curve

$$y = x^p(x+1)^q(x-1.5)^r.$$

a. Use the sketch to find the integers p, q, and r. Check by generating the curve with x-min $= -2$, x-max $= 3$, x-scale $= 1$, y-min $= -5$, y-max $= 6$, and y-scale $= 1$.

b. Explain how the values of p, q, and r determine the shape of the curve.

Problem 4a   Find integers p, q, and r such that the curve in Figure 4 is the graph of

$$y = x^p(x-0.5)^q(x-1)^r.$$

Check by generating the curve with x-min $= -2$, x-max $= 3$, x-scale $= 1$, y-min $= -4$, y-max $= 4$, and y-scale $= 1$.

b. How do these values of p, q, and r determine the shape of the curve?

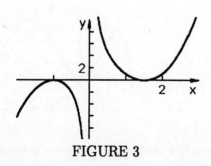

FIGURE 3

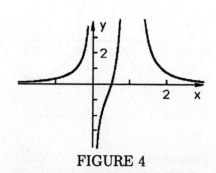

FIGURE 4

Graphing calculator workbook

Worksheet 2F.3[†]

Fractional and negative powers

Name _____ Date _____

Others in your group _____

Instructor, Teaching Assistant, and/or Recitation section _____

Directions Put your initial calculations and first drafts of your conclusions on scratch paper. Take your time, work carefully, and give your conclusions in well written sentences. Discuss your solution with at least one other student before putting a final draft on this sheet or other sheets of paper, as needed. Turn in all your work.

Problem 1 Draw the line $y = x$ and the curve $y = x^3$ in Figure 1 by plotting the points at $x = 0, \pm 0.5, \pm 1,$ and ± 1.2. Check your drawing with your calculator or computer.

Explain why changing $y = x$ to $y = x^3$ changes the graph as it does. (Hint: How do x and x^3 compare for $x < -1, -1 < x < 0, 0 < x < 1,$ and $x > 1$?)

Problem 2 Draw the line $y = x$ and the curve $y = x^{1/3}$ in Figure 2 by plotting the points at $x = 0, \pm 0.5, \pm 1$ and ± 1.5. Check your drawing with your calculator or computer.

Explain why changing $y = x$ to $y = x^{1/3}$ changes the graph as it does. (Hint: How do x and $x^{1/3}$ compare for $x < -1, -1 < x < 0, 0 < x < 1,$ and $x > 1$?)

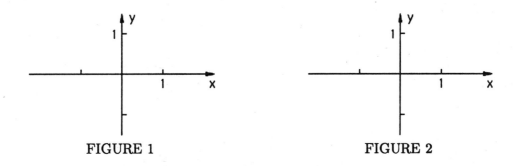

FIGURE 1   FIGURE 2

[†]This worksheet does not use any calculus. It involves analyzing the shape of $y = x^n$ for fractional n.

Problem 3 Draw the curves $y = \dfrac{1}{x^2}$ and $y = \left(\dfrac{1}{x^2}\right)^2$ in Figure 3 by plotting the points at $x = \pm 0.75, \pm 1$, and ± 2. Check your drawing with your calculator or computer.

Explain why changing $y = \dfrac{1}{x^2}$ to $y = \left(\dfrac{1}{x^2}\right)^2$ changes the graph as it does.

Problem 4 Draw the curves $y = \dfrac{1}{x^2}$ and $y = \left(\dfrac{1}{x^2}\right)^{1/4}$ in Figure 4 by plotting the points at $x = \pm 0.75, \pm 1$, and ± 2. Check your drawing with your calculator or computer.

Explain why changing $y = \dfrac{1}{x^2}$ to $y = \left(\dfrac{1}{x^2}\right)^{1/4}$ changes the graph as it does.

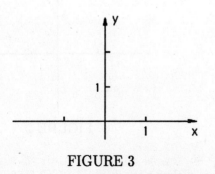

FIGURE 3

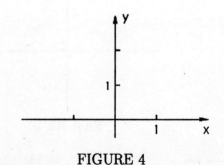

FIGURE 4

Graphing calculator workbook

Worksheet 2G.1[†]

The first-derivative test: Initial Investigations

Name _____ Date _____

Others in your group _____

Instructor, Teaching Assistant, and/or Recitation section _____

Directions Put first drafts of your calculations and answers on scratch paper. Take your time, work carefully, and discuss your solution with at least one other student before putting a final draft on this sheet or on other paper. Turn in all your work.

Problem 1 Generate the graph of $f(x) = x^4 - 4x^3 + 4x^2 - 2$ for $-1 \leq x \leq 3, -3 \leq y \leq 3$ on your calculator or computer and copy it in Figure 1.

a. $f(x)$ is increasing (as x increases) in two open intervals with integer or infinite endpoints and is decreasing in two open intervals with integer or infinite endpoints. Use the graph to find those intervals.

b. Find the derivative $f'(x)$ of the function $f(x)$ of part (a). Generate its graph for $-1 \leq x \leq 3, -8 \leq y \leq 8$ and compare the result with Figure 2 to check your calculations.

c. $f'(x)$ is positive in two open intervals with integer or infinite endpoints and negative in two open intervals with integer or infinite endpoints. Use Figure 2 to determine those intervals.

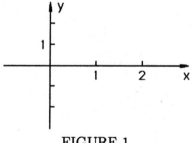
FIGURE 1

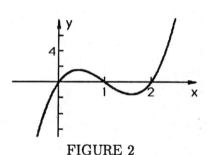
FIGURE 2

Problem 2 Generate the graph of $g(x) = \dfrac{x}{x^2 + 4}$ for $-8 \leq x \leq 8, -0.4 \leq y \leq 0.4$ on your calculator or computer and copy it in Figure 3.

a. $g(x)$ is increasing or decreasing in open intervals with integer or infinite endpoints. Use the graph to find those intervals.

b. Find the derivative $g'(x)$ of the function $g(x)$ of part (a). Generate its graph for $-8 \leq x \leq 8, -0.1 \leq y \leq 0.4$ and compare the result with Figure 4 to check your work.

[†]This worksheet can be used to introduce or illustrate the first derivative test. It uses the rules for differentiating powers, linear combinations, and quotients.

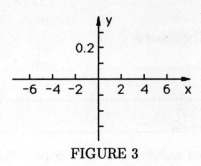
FIGURE 3

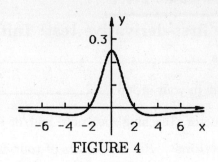
FIGURE 4

c. $g'(x)$ is positive or negative in open intervals with integer or infinite endpoints. Use Figure 4 to determine those intervals.

Problem 3 Generate the graph of $h(x) = \dfrac{1}{x} + \dfrac{1}{3}x^3$ for $-2.5 \le x \le 2.5, -6 \le y \le 6$ on your calculator or computer and copy it in Figure 5.

a. $h(x)$ is defined and increasing or decreasing in open intervals with integer or infinite endpoints. Use the graph to find those intervals. (Be sure to include 0 as an endpoint of two of the intervals.)

b. Find the derivative $h'(x)$ of the function $h(x)$ of part (a). Generate its graph with the ranges from part (a) and compare the result with Figure 6 to check your work.

c. $h'(x)$ is defined and positive or negative in open intervals with integer or infinite endpoints. Use Figure 6 to determine those intervals.

Problem 4 What is the pattern in the answers to parts (a) and (c) of Problems 1, 2, and 3? Explain.

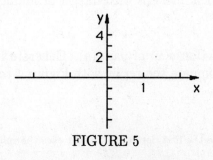
FIGURE 5

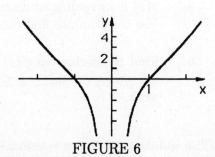
FIGURE 6

Graphing calculator workbook

Worksheet 2G.2†

The second-derivative test: Initial Investigations

Name _____ Date _____

Others in your group _____

Instructor, Teaching Assistant, and/or Recitation section _____

Directions Put first drafts of your calculations and answers on scratch paper. Take your time, work carefully, and discuss your solution with at least one other student before putting a final draft on this sheet or on other paper. Turn in all your work.

Problem 1 Generate the graph of $F(x) = \frac{2}{3}x^4 - 4x^2$ for $-3 \le x \le 3, -8 \le y \le 8$ on your calculator or computer and copy it in Figure 1.

  a. The graph of $F(x)$ is concave up in two open intervals with integer or infinite endpoints and is concave down in one open interval with integer or infinite endpoints. Use Figure 1 to find those intervals.

  b. Find the second derivative $F''(x)$ of the function $F(x)$ of part (a). Generate its graph for $-3 \le x \le 3, -15 \le y \le 30$ and compare the result with Figure 2 to check your calculations.

  c. $F''(x)$ is positive in two open intervals with integer or infinite endpoints and negative in one open interval with integer or infinite endpoints. Use Figure 2 to find those intervals.

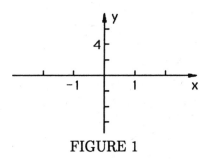
FIGURE 1

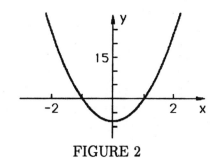
FIGURE 2

Problem 2 Generate the graph of $G(x) = 10x + \dfrac{1}{x^2} - x^3$ for $-4 \le x \le 4, -20 \le y \le 40$ with y-scale $= 10$ and copy it in Figure 3.

  a. The graph of $G(x)$ is concave up in two open intervals with integer or infinite endpoints and concave down in one open interval with integer or infinite endpoints. Use Figure 3 to find those intervals.

  b. Find the second derivative $G''(x)$ of the function $G(x)$ of part (a). Generate its graph with the ranges of x and y from part (a) and compare the result with Figure 4 to check your work.

---

†This worksheet can be used to introduce or illustrate the second derivative test. It uses the chain rule and rules for differentiating powers, linear combinations, and quotients.

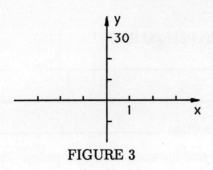

FIGURE 3

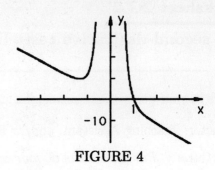

FIGURE 4

c. $G''(x)$ is defined and positive or negative in open intervals with integer or infinite endpoints. Use Figure 4 to determine those intervals.

Problem 3 What is the pattern in the answers to parts (a) and (c) of Problems 1 and 2? Explain.

Problem 4 Generate the graph of $H(x) = x - \dfrac{1}{x^2 + 12}$ for $-8 \leq x \leq 8$, $-6 \leq y \leq 6$ and copy it in Figure 5. The graph looks like a line but is not.

a. Find the second derivative $H''(x)$ of the function $H(x)$ of part (a) (a long calculation). Use your formula to generate its graph for $-8 \leq x \leq 8, -0.01 \leq y \leq 0.02$ and copy it in Figure 6.

b. $H''(x)$ is positive or negative in open intervals with integer or infinite endpoints. Use your sketch of its graph to determine those intervals. Then use the principle predicted in Problem 3 to find the open intervals where the graph of $H(x)$ is concave up and where it is concave down.

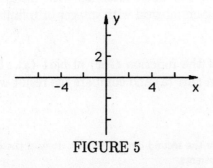

FIGURE 5

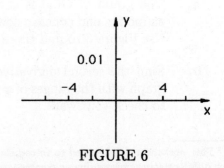

FIGURE 6

Graphing calculator workbook   101

Worksheet 2G.3[†]

Graphs of functions constructed from the sine and cosine

Name _____ Date _____

Others in your group _____

Instructor, Teaching Assistant, and/or Recitation section _____

Directions   Put first drafts of your calculations and answers on scratch paper. Take your time, work carefully, and discuss your solution with at least one other student before putting a final draft on this sheet or on other paper. Turn in all your work.

Problem 1   Generate the curve $y = \dfrac{\cos(\pi x)}{x^2 + 1}$ on your calculator or computer for $-3 \le x \le 3$, $-1.2 \le y \le 1.5$ and copy it in Figure 1. Explain its shape without using any calculus. (Hint: Regenerate it with the curves $y = \dfrac{\pm 1}{x^2 + 1}$.)

Problem 2   Generate $y = \cos(\pi x) + \frac{1}{9}x^2$ for $-4 \le x \le 4$, $-1.5 \le y \le 3$ and copy it in Figure 2. Explain its shape without using any calculus. (Hint: Regenerate it with the curves $y = \frac{1}{9}x^2 \pm 1$.)

Problem 3   Generate the curves $y = \sin^2 x + \cos^2 x$, $y = \sec^2 x - \tan^2 x$, and $y = \csc^2 x - \cot^2 x$, for $-6 \le x \le 6$, $-0.5 \le y \le 1.5$ and copy them in Figure 3. Enter $\sin^2 x$ as $(\sin x)^2$, etc. The graphs look the same but in fact represent different functions—functions with different domains. Explain.

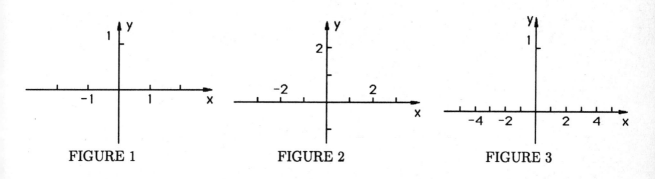

FIGURE 1   FIGURE 2   FIGURE 3

[†]These problems call for analyzing calculator- or computer-generated graphs by using properties of the sine and cosine. They do not require any calculus.

Problem 4a Generate the graph of $\cos^2 x$ with the graph of $\frac{1}{2}\cos(2x)$ for $-6 \le x \le 6$, $-1.2 \le y \le 1.5$ and copy them in Figure 4. These curves illustrate a trigonometric identity. What is it?

Problem 5a Generate the graph of $\sqrt{\sin(\pi x)}$ for $-2 \le x \le 3$, $-1.2 \le y \le 1.5$ and copy it in Figure 5. Explain its shape without using any calculus. (Hint: Regenerate it with the graph of $\sin(\pi x)$, and consider what taking square roots does to negative numbers, numbers between 0 and 1, and numbers greater than 1.)

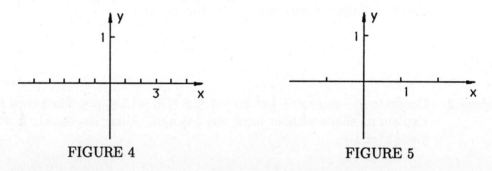

FIGURE 4 FIGURE 5

b. Generate the graph of $(\sin(\pi x))^3$ with the ranges from part (a) and copy it in Figure 6. Explain its shape without using any calculus. (Hint: Compare it with the graph of $\sin(\pi x)$.)

c. Generate the graph of $(\sin(\pi x))^{1/3}$ with the ranges from part (a) and copy it in Figure 7. Explain its shape without using any calculus. (Hint: Compare it with the graph of $\sin(\pi x)$.)

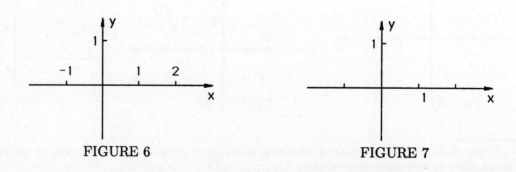

FIGURE 6 FIGURE 7

Graphing calculator workbook

Worksheet 2G.4[†]

Graphs with vertical tangent lines, cusps, and limited extent

Name _____ Date _____

Others in your group _____

Instructor, Teaching Assistant, and/or Recitation section _____

Directions   Put first drafts of your calculations and answers on scratch paper. Take your time, work carefully, and discuss your solution with at least one other student before putting a final draft on this sheet or on other paper. Turn in all your work.

Problem 1   Generate the graph of $(x^2 - 3x)^{1/3}$ on a calculator or computer for $-1 \leq x \leq 4$, $-1.5 \leq y \leq 2$ and copy it in Figure 1. Use the first-derivative to find the open intervals where the function is increasing or decreasing and to show that the graph has two vertical tangent lines.

Problem 2   Generate the graph of $(x^2 - 1)^{2/3}$ for $-2 \leq x \leq 2$, $-0.5 \leq y \leq 2.7$ and copy it in Figure 2. Enter $(x^2 - 1)^{2/3}$ as $((x^2 - 1)^2)^{1/3}$. Use the first-derivative to find the open intervals where the function is increasing or decreasing and to show that the graph has two cusps.

Problem 3   Generate the graph of $\frac{1}{2}x^2 - 3x^{1/3}$ for $-2 \leq x \leq 5$, $-3 \leq y \leq 7$ and copy it in Figure 3. Use the first-derivative to find the open intervals where the function is increasing or decreasing and to show that the graph has a vertical tangent line.

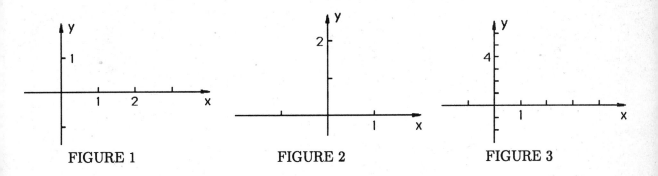

FIGURE 1   FIGURE 2   FIGURE 3

[†]These problems provide practice analyzing calculator- or computer-generated graphs by studying derivatives.

Problem 4 Generate the graph of $(x^2 - 1)^{1/4}$ for $-5 \leq x \leq 5$, $-0.5 \leq y \leq 2.5$ and copy it in Figure 4. Use the formulas for the function and for its derivative to explain the shape of the curve. The two branches of the graph extend down to the x-axis. The graph might be said to have two one-sided vertical tangent lines. Explain.

Problem 5 Generate the graph of $(81 - x^4)^{1/4}$ for $-5 \leq x \leq 5$, $-1.67 \leq y \leq 5$ and copy it in Figure 5. It is half of what has been called a "super circle." Describe how it differs from the semcircle $y = (9 - x^2)^{1/2}$ and why. (Show that $81 - x^4$ is greater than $(9 - x^2)^2$ for $0 < |x| < 3$.)

Problem 6 Generate $y = \dfrac{1}{\sqrt{4-x}}$ for $-2 \leq x \leq 5$, $-1 \leq y \leq 6$ and copy it in Figure 6. Explain its shape by studying the function and its first- and second derivatives.

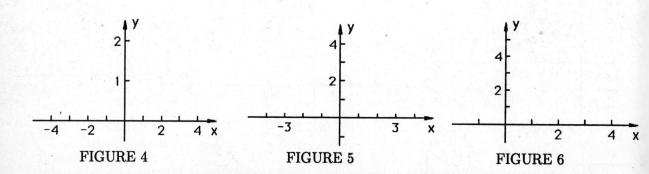

FIGURE 4   FIGURE 5   FIGURE 6

Graphing calculator workbook

Worksheet 2H.1†

Rates of change of functions whose graphs are lines

Name _____ Date _____

Others in your group _____

Instructor, Teaching Assistant, and/or Recitation section _____

Directions Put your initial calculations and first drafts of your conclusions on scratch paper. Take your time, work carefully, and give your conclusions in well written sentences. Discuss your solution with at least one other student before putting a final draft on this sheet or other sheets of paper, as needed. Turn in all your work.

Problem 1 Water freezes at 0 degrees Celsius and boils at 100 degrees Celsius. It freezes at 32 degrees Fahrenheit and boils at 212 degrees Fahrenheit.

a. Which of Figures 1 through 3 is the graph of the temperature y, measured in degrees Fahrenheit, as a function of the temperature x, measured in degrees Celsius? Make your choice using as little algebra as possible and explain your reasoning.

b. Give a formula for degrees Fahrenheit y as a function of degrees Celsius x. Generate its graph on your calculator or computer and compare it with the figure you chose in part (a) to check your answer. Use x-min $= -10$, x-max $= 110$, x-scale $= 25$, y-min $= -100$, y-max $= 250$, and y-scale $= 25$.

c. Find $\dfrac{dy}{dx}$ from your formula in part (b). Explain its physical significance and how it relates to the line from part (a).

d. Solve for x as a function of y. Then find $\dfrac{dx}{dy}$ and explain its physical significance.

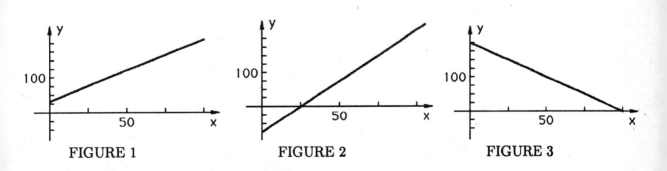

FIGURE 1   FIGURE 2   FIGURE 3

---

†These problems study rates of change of linear functions.

106   Graphing calculator workbook

Problem 2   You want to spend exactly \$600 on two kinds of cheese by purchasing x pounds
of Gorgonzola cheese, which costs \$6 per pound, and y pounds of Monterey jack
cheese, which costs \$3 per pound.

   a.   Which of Figures 1 through 3 shows the line relating possible values of x and y?
Explain how you make your choice.

   b.   Give the pounds y of Monterey jack cheese you would buy as a function of the
pounds x of Gorgonzola that you purchase. Generate its graph with the ranges
of x and y from Problem 1. Compare it with the line from part (a) to check your
answers.

   c.   Find $\dfrac{dy}{dx}$. Explain its economic significance and how it is related to the corresponding line.

   d.   Solve for x as a function of y. Then find $\dfrac{dx}{dy}$ and explain its economic significance.

Problem 3   Georgia's Generic Business requires \$75 each day in overhead and salaries. Each
unit of her Generic Product costs an additional \$2 to produce and sells for \$5.

   a.   Which of Figures 1 through 3 shows Georgia's profit y as a function of the number
x of units she makes in a day, under the assumption that she sells all she produces?
Explain your choice.

   b.   Express the profit y as a function of x. Generate its graph with the ranges of x
and y from Problem 1 and compare it with the line from part (a).

   c.   Find $\dfrac{dy}{dx}$. Then explain its economic significance and how it relates to the line from
part (a).

Graphing calculator workbook

Worksheet 2H.2[†]

The rate of change of one distance with respect to another

Name _____ Date _____

Others in your group _____

Instructor, Teaching Assistant, and/or Recitation section _____

Directions Put your initial calculations and first drafts of your conclusions on scratch paper. Take your time, work carefully, and give your conclusions in well written sentences. Discuss your solution with at least one other student before putting a final draft on this sheet or other sheets of paper, as needed. Turn in all your work.

Problem 1 Daisy drives three miles south from her home to the intersection of her road with an east-west highway. Then she turns left and drives east (Figure 1).

a. Give the distance L (miles) from her house as a function of her distance x (miles) from the intersection. Generate the graph of this function on your calculator or computer, using y in place of L and with $-1 \leq x \leq 8, -2 \leq y \leq 8$. Then generate the line $L = x$ on the same screen. Copy the curve and the line in Figure 2.

b. How far is she from the intersection when she is 6 miles from her house? Give the exact answer and its approximate decimal value. Check by using the trace command on your calculator or the graph on your computer.

c. Explain, using the geometric relationships in Figure 1, why $L = x$ is an asymptote of the curve in Figure 2. (Hint: The arc IP of the circle with its center at C in Figure 3 marks off the distance x on the line CH. Draw this triangle with the same height but a much larger value of x to see how increasing x changes the triangle. Then draw the arc through I with center at C in the larger triangle.)

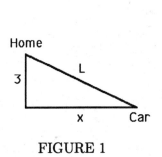

FIGURE 1

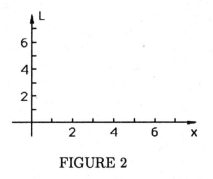

FIGURE 2

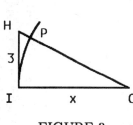

FIGURE 3

[†]This worksheet explores a narrative problem dealing with rates of change.

d. Give $\dfrac{dL}{dx}$ as a function of x for the function L of Problem 1. Generate the graph of this derivative with y in place of $\dfrac{dL}{dx}$, $-1 \le x \le 8, -0.2 \le y \le 1.3$ and y-scale $= 0.1$. Copy the graph in Figure 4.

e. How far does she have to go past the intersection before $\dfrac{dL}{dx} = 0.90$? Give the exact answer and its approximate decimal value. Check it with the trace function on your calculator or the graph on your computer.

f. What happens to $\dfrac{dL}{dx}$ as $x \to \infty$? Verify this algebraically. (Write $\dfrac{dL}{dx}$ as $\dfrac{x}{\sqrt{g(x)}}$ for some function $g(x)$. Then for $x > 0$, divide the numerator by x and the denominator by $\sqrt{x^2} = x$.)

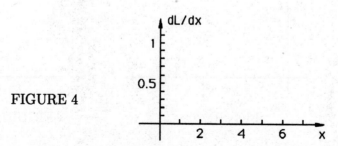

FIGURE 4

Graphing calculator workbook

Worksheet 2H.3†

Rates of change of the volume of a tent

Name _____ Date _____

Others in your group _____

Instructor, Teaching Assistant, and/or Recitation section _____

Directions Put your initial calculations and first drafts of your conclusions on scratch paper. Take your time, work carefully, and give your conclusions in well written sentences. Discuss your solution with at least one other student before putting a final draft on this sheet or other sheets of paper, as needed. Turn in all your work.

A tent is to be constructed in the shape of a right circular cone as in Figure 1. Its volume is $V = \frac{1}{3}\pi r^2 h$, where r is radius of its base and h is its height.

Problem 1 Suppose that the radius is held at the constant value $r = 3$ while the height h varies.

a. Express the volume V as a function of h. Generate its graph on your calculator or computer, using x in place of h and y in place of V and with $-1 \leq x \leq 6$, $-5 \leq y \leq 60$, y-scale $= 10$. Copy the graph in Figure 2.

b. How does doubling the height affect the volume of the tent?

c. Find the rate of change of the volume with respect to the height. Generate the graph of this function with x in place of h, y in place of $\frac{dV}{dh}$, $-1 \leq x \leq 6, -5 \leq y \leq 25$ and y-scale $= 5$. Copy the graph in Figure 3.

Problem 2 Suppose that, instead of the conditions of Problem 1, the height has the constant value $h = 2$ and it is r that varies.

a. Express V as a function of r. Generate its graph with y in place of V, x in place of r, and the range from Problem 1a. Copy it in Figure 4.

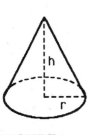

FIGURE 1

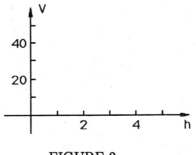

FIGURE 2

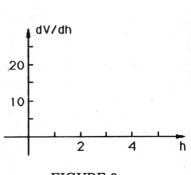

FIGURE 3

---

†This worksheet explores a narrative problem dealing with rates of change.

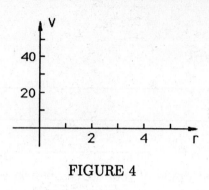

FIGURE 4

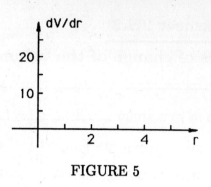

FIGURE 5

- b. How does doubling the radius affect the volume in this case?

- c. Find the rate of change of V with respect to r. Generate its graph with y in place of dV/dr, x in place of r, and the ranges from Problem 1c. Copy it in Figure 5.

Problem 3 Now suppose that the radius of the base and height of the tent both vary with the radius always equal to half the height.

- a. Express V as a function of h. Generate its graph with the ranges from Problem 1a and copy it in Figure 6.

- b. How does doubling the height affect the volume in this case?

- c. Find $\dfrac{dV}{dh}$ as a function of h. Generate its graph with the ranges from Problem 1c and copy it in Figure 7.

Problem 4 Find a way to have r be a function of h so that doubling h multiplies V by 32.

Problem 5 Problems 1, 3, and 4 illustrate how the rate of change $\dfrac{dV}{dh}$ of the volume of the tent with respect to its height can have different meanings in different contexts. Find a way for r to be a function of h so that $\dfrac{dV}{dh}$ equals the constant 0.

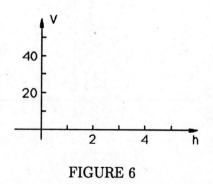

FIGURE 6

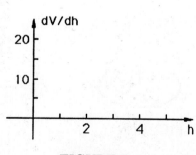

FIGURE 7

Graphing calculator workbook

Worksheet 2H.4[†]

A rate of change of the height of a ladder

Name _____ Date _____

Others in your group _____

Instructor, Teaching Assistant, and/or Recitation section _____

Directions Put your initial calculations and first drafts of your conclusions on scratch paper. Take your time, work carefully, and give your conclusions in well written sentences. Discuss your solution with at least one other student before putting a final draft on this sheet or other sheets of paper, as needed. Turn in all your work.

Problem 1 A 10–foot–long ladder is leaning against a vertical wall with its base on the horizontal ground (Figure 1).

  a. Give the height h of the top of the ladder above the ground as a function of the distance s of its base from the bottom of the wall. Generate the graph of this function on your calculator or computer, with x in place of s, y in place of h, $-1 \leq x \leq 12$, x-scale = 2, $-1 \leq y \leq 12$, and y-scale = 2. Copy the portion for $s \geq 0$ in Figure 2.

  b. Use the geometric relationships in Figure 1 to explain why $h = 10$ when $s = 0$ and $h = 0$ when $s = 10$.

  c. The graph is one-fourth of an ellipse. What would it be if the scales were equal on the s- and h-axes?

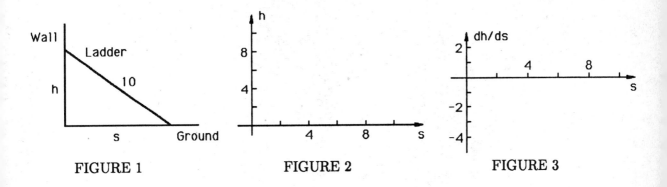

FIGURE 1   FIGURE 2   FIGURE 3

[†]This worksheet explores a narrative problem dealing with rates of change.

d. What is the rate of change of h with respect to s? Generate the graph of this function with x in place of s, y in place of $\dfrac{dh}{ds}$, $-1 \le x \le 12$, x-scale $= 2$, and $-6 \le y \le 3$. Copy the graph in Figure 3.

e. Give a geometric explanation, based on Figure 1, of why $\dfrac{dh}{ds}$ is negative.

f. What is $\lim\limits_{s \to 0^+} \dfrac{dh}{ds}$? Explain what this means geometrically by considering Figure 1 for small, positive values of s.

g. What is $\lim\limits_{s \to 10^-} \dfrac{dh}{ds}$? Explain what this means geometrically by considering Figure 1 for values of s that are slightly less than 10.

h. Suppose the base of the ladder is being pulled away from the wall at a constant speed of 1 foot per hour. By the chain rule, the upward velocity of the top of the ladder is

$$\frac{dh}{dt} = \frac{dh}{ds}\frac{ds}{dt} = \frac{dh}{ds}.$$

What happens as $s \to 10^-$, according to the result of part (g)? What probably really happens?

Graphing calculator workbook   113

Worksheet 2H.5[†]

A rate of change of the length of a shadow

Name _____ Date _____

Others in your group _____

Instructor, Teaching Assistant, and/or Recitation section _____

Directions   Put your initial calculations and first drafts of your conclusions on scratch paper. Take your time, work carefully, and give your conclusions in well written sentences. Discuss your solution with at least one other student before putting a final draft on this sheet or other sheets of paper, as needed. Turn in all your work.

Problem 1   A light on the horizontal ground is casting the shadow of a 10–foot–high fence on the side of a vertical skyscraper (Figure 1). The fence is 8 feet from the skyscraper.

    a.   How high does the shadow of the fence reach on the side of the skyscraper if the lamp is eight feet from the base of the fence as in Figure 2? (Use similar triangles.)

    b.   How high does the shadow reach if the lamp is 16 feet from the base of the fence as in Figure 3?

    c.   Give the height h of the shadow as a function of the distance s of the lamp from the fence. (Use Figure 1 and check your formula with the results of parts (a) and (b).) Generate its graph on your calculator or computer, using s in place of x, y in place of h, $-6 \leq x \leq 24$, x-scale $= 5$, $-4 \leq y \leq 50$, and y-scale $= 5$. Copy the graph in Figure 4.

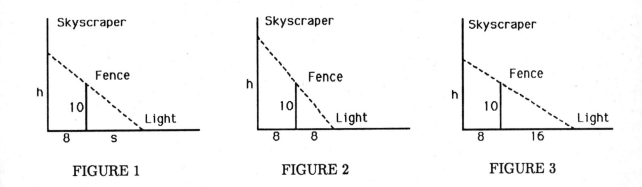

    FIGURE 1   FIGURE 2   FIGURE 3

[†]This worksheet explores a narrative problem dealing with rates of change.

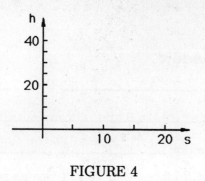
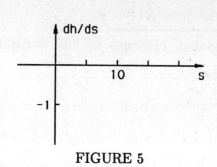

FIGURE 4   FIGURE 5

d.   What are $\lim\limits_{s \to 0^+} h$ and $\lim\limits_{s \to \infty} h$?
Use the geometric relationships in Figure 1 to explain these results.

e.   Find the rate of change of the height of the shadow with respect to the distance s from the lamp to the fence. Generate its graph with s in place of x, y in place of dh/ds, $-6 \leq x \leq 24$, x-scale $= 5$, and $-2 \leq y \leq 1$. Copy it in Figure 5.

f.   What are $\lim\limits_{s \to 0^+} \dfrac{dh}{ds}$ and $\lim\limits_{s \to \infty} \dfrac{dh}{ds}$?

g.   Give geometric interpretations of the results of part (f).

Graphing calculator workbook

Newton's method program 2I

Texas Instruments TI-81

Newton's method calculates approximate solutions of equations of the form $F(x) = 0$ by finding x-intercepts of tangent lines to the graph of $F(x)$. An initial approximate solution x_0 is given by the user. Then x_1 is the x-intercept of the tangent line at $x = x_0$; x_2 is the x-intercept of the tangent line at $x = x_1$; and so forth. In general, x_{j+1} is obtained from x_j by

$$(1) \qquad x_{j+1} = x_j - \frac{F(x_j)}{F'(x_j)} .$$

In most cases when the method works, the numbers $x_0, x_1, x_2, \ldots$ approach a solution of $F(x) = 0$ quickly, and after a certain point, successive values are equal in all the digits displayed by the calculator or computer. The repeated value of x is taken as the approximate solution.

Instructions are given here for entering the following program that draws the graph and the tangent lines as the calculations in Newton's method are carried out with $F(X) = Y_1(X)$ and $F'(X) = Y_2(X)$. The line numbers are given only for reference and are not used on the calculator. Zero is given as $\emptyset$ to distinguish it from the letter O.

The program (172 bytes)

1.   Prgm2: NEWTON
2.   : ClrDraw
3.   : All-Off
4.   : Y_1-On
5.   : Disp "SOLVE $Y_1(X) = \emptyset$ "
6.   : Disp "SET $Y_2 = DY_1$ "
7.   : Disp "$X\emptyset =$ "
8.   : Input Z
9.   : $\emptyset \to J$
10.   : Lbl 1
11.   : $Z \to X$
12.   : $Y_1 \to W$
13.   : $Y_2 \to M$
14.   : Line$(Z, \emptyset, Z, W)$
15.   : Pause
16.   : Disp "$J, XJ, F(XJ) =$ "
17.   : Disp J
18.   : Disp Z
19.   : Disp W
20.   : Pause
21.   : DrawF $W + M(X - Z)$
22.   : If $M = \emptyset$
23.   : Goto 2

24.   : $Z - W/M \to Z$

25.   : $J + 1 \to J$

26.   : Goto 1

27.   : Lbl 2

28.   : Pause

29.   : Disp "SLOPE = $\emptyset$"

30.   : Disp "THE METHOD FAILS"

Entering the program

Press $\boxed{\text{2nd}}$ $\boxed{\text{QUIT}}$ to display the home screen. Then press $\boxed{\text{PRGM}}$ for the menu of programs. Press $\boxed{\blacktriangleright}$ so that EDIT is highlighted and $\boxed{2}$ to use program #2 or another number if program #2 has already been used. If you make a mistake, move the cursor to the error and use $\boxed{\text{INS}}$ and $\boxed{\text{DEL}}$ to correct it. Press $\boxed{\text{QUIT}}$ or follow the steps at the beginning of this paragraph to return to the program from other screens. Press $\boxed{\text{ON}}$ $\boxed{2}$ (QUIT) to interrupt the running of the program.

---

1. Press $\boxed{\text{N}}$ $\boxed{\text{E}}$ $\boxed{\text{W}}$ $\boxed{\text{T}}$ $\boxed{\text{O}}$ $\boxed{\text{N}}$ $\boxed{\text{ENTER}}$. This title identifies the program in the program menu.

---

2. Press $\boxed{\text{2nd}}$ $\boxed{\text{DRAW}}$ $\boxed{1}$ $\boxed{\text{ENTER}}$. ClrDraw clears the graphics screen when the program is run.

---

3. Press $\boxed{\text{2nd}}$ $\boxed{\text{Y-VARS}}$ $\boxed{\blacktriangleleft}$ $\boxed{1}$ $\boxed{\text{ENTER}}$. All-Off unselects all the formulas in the Y= menu so their graphs will not be generated when the program is run.

---

4. Press $\boxed{\text{2nd}}$ $\boxed{\text{Y-VARS}}$ $\boxed{\blacktriangleright}$ $\boxed{2}$ $\boxed{\text{ENTER}}$. Y_1-On selects Y_1 so its graph will be generated.

---

5. Press $\boxed{\text{PRGM}}$ to display CTL (control) commands and then $\boxed{\blacktriangleright}$ for I/O (input/output) commands. Press $\boxed{1}$ to enter the word Disp. Press $\boxed{\text{2nd}}$ $\boxed{\text{A-LOCK}}$ to put the calculator in alpha mode and $\boxed{\text{``}}$ $\boxed{\text{S}}$ $\boxed{\text{O}}$ $\boxed{\text{L}}$ $\boxed{\text{V}}$ $\boxed{\text{E}}$ $\boxed{\text{-}}$ $\boxed{\text{2nd}}$ $\boxed{\text{Y-VARS}}$ $\boxed{1}$ $\boxed{(}$ $\boxed{X|T}$ $\boxed{)}$ $\boxed{\text{2nd}}$ $\boxed{\text{TEST}}$ $\boxed{1}$ $\boxed{\emptyset}$ $\boxed{\text{ALPHA}}$ $\boxed{\text{``}}$ $\boxed{\text{ENTER}}$ with $\boxed{\text{-}}$ the space on the zero key.

---

6. Press $\boxed{\text{PRGM}}$ $\boxed{\blacktriangleright}$ $\boxed{1}$ $\boxed{\text{2nd}}$ $\boxed{\text{A-LOCK}}$ $\boxed{\text{``}}$ $\boxed{\text{S}}$ $\boxed{\text{E}}$ $\boxed{\text{T}}$ $\boxed{\text{-}}$ $\boxed{\text{2nd}}$ $\boxed{\text{Y-VARS}}$ $\boxed{2}$ $\boxed{\text{2nd}}$ $\boxed{\text{TEST}}$ $\boxed{1}$ $\boxed{\text{ALPHA}}$ $\boxed{\text{D}}$ $\boxed{\text{2nd}}$ $\boxed{\text{Y-VARS}}$ $\boxed{1}$ $\boxed{\text{ALPHA}}$ $\boxed{\text{``}}$ $\boxed{\text{ENTER}}$. This and the previous lines are reminders that Y_1 should be F and Y_2 should be its derivative.

Graphing calculator workbook

7. Press $\boxed{\text{PRGM}}$ $\boxed{\blacktriangleright}$ $\boxed{1}$ $\boxed{\text{ALPHA}}$ $\boxed{``}$ $\boxed{X|T}$ $\boxed{0}$ $\boxed{\text{2nd}}$ $\boxed{\text{TEST}}$
$\boxed{1}$ $\boxed{\text{ALPHA}}$ $\boxed{``}$ $\boxed{\text{ENTER}}$ for the prompt $X0 =?$.

8. Press $\boxed{\text{PRGM}}$ $\boxed{\blacktriangleright}$ $\boxed{2}$ $\boxed{\text{ALPHA}}$ $\boxed{Z}$ $\boxed{\text{ENTER}}$. Here the program pauses for the user to assign a value to $X0$, which is stored as Z.

9. Press $\boxed{0}$ $\boxed{\text{STO}\blacktriangleright}$ $\boxed{J}$ $\boxed{\text{ENTER}}$. J counts the number of iterations of Newton's method that have been performed. It is set equal to 0 since we are at the beginning of the program.

10. Press $\boxed{\text{PRGM}}$ $\boxed{1}$ $\boxed{1}$ $\boxed{\text{ENTER}}$. The operation of the program will return to this label when it reaches the Goto 1 that will be put on line 26.

11. Press $\boxed{\text{ALPHA}}$ $\boxed{Z}$ $\boxed{\text{STO}\blacktriangleright}$ $\boxed{X|T}$ $\boxed{\text{ENTER}}$. This assigns the value of Z to X.

12. Press $\boxed{\text{2nd}}$ $\boxed{\text{Y-vars}}$ $\boxed{1}$ $\boxed{\text{STO}\blacktriangleright}$ $\boxed{W}$ $\boxed{\text{ENTER}}$. The value at $X = Z$ of the function $Y_1(= F(x))$ in the Y= menu is stored as W.

13. Press $\boxed{\text{2nd}}$ $\boxed{\text{Y-vars}}$ $\boxed{2}$ $\boxed{\text{STO}\blacktriangleright}$ $\boxed{M}$ $\boxed{\text{ENTER}}$. The value of the derivative $Y_2 = F'(x)$ at $X(= Z)$ is stored as M.

14. Press $\boxed{\text{2nd}}$ $\boxed{\text{DRAW}}$ $\boxed{2}$ $\boxed{\text{ALPHA}}$ $\boxed{Z}$ $\boxed{\text{ALPHA}}$ $\boxed{,}$ $\boxed{0}$ $\boxed{\text{ALPHA}}$
$\boxed{,}$ $\boxed{\text{ALPHA}}$ $\boxed{Z}$ $\boxed{\text{ALPHA}}$ $\boxed{,}$ $\boxed{\text{ALPHA}}$ $\boxed{W}$ $\boxed{)}$ $\boxed{\text{ENTER}}$.
This draws a vertical line from $(Z,0) = (XJ,0)$ on the x-axis to $(Z,W) = (XJ, F(XJ))$ on the graph of $F(x)$.

15. Press $\boxed{\text{PRGM}}$ $\boxed{6}$ $\boxed{\text{ENTER}}$ to insert a PAUSE. This command stops the program so the user can see the graphs.

16. Press $\boxed{\text{PRGM}}$ $\boxed{\blacktriangleright}$ $\boxed{1}$ $\boxed{\text{2nd}}$ $\boxed{\text{A-LOCK}}$ $\boxed{``}$ $\boxed{J}$ $\boxed{,}$ $\boxed{X|T}$ $\boxed{J}$
$\boxed{,}$ $\boxed{F}$. Press $\boxed{\text{ALPHA}}$ to leave alpha mode and then $\boxed{(}$ $\boxed{X|T}$ $\boxed{\text{ALPHA}}$
$\boxed{J}$ $\boxed{)}$ $\boxed{\text{2nd}}$ $\boxed{\text{TEST}}$ $\boxed{1}$ $\boxed{\text{ALPHA}}$ $\boxed{``}$ $\boxed{\text{ENTER}}$. The symbols
$J, XJ, F(XJ) =$ will appear in the home screen when the program is run.

17. Press $\boxed{\text{PRGM}}$ $\boxed{\blacktriangleright}$ $\boxed{1}$ $\boxed{\text{ALPHA}}$ $\boxed{J}$ $\boxed{\text{ENTER}}$. The value of J is displayed.

18. Press $\boxed{\text{PRGM}}$ $\boxed{\blacktriangleright}$ $\boxed{1}$ $\boxed{\text{ALPHA}}$ $\boxed{Z}$ $\boxed{\text{ENTER}}$. The value of $Z = XJ$ is displayed.

19. Press $\boxed{\text{PRGM}}$ $\boxed{\blacktriangleright}$ $\boxed{1}$ $\boxed{\text{ALPHA}}$ $\boxed{W}$ $\boxed{\text{ENTER}}$. The value of $W = F(XJ)$ is displayed.

20. Press $\boxed{\text{PRGM}}$ $\boxed{6}$ $\boxed{\text{ENTER}}$ to insert a PAUSE so the user can read the values of J, XJ, and $F(XJ)$.

21. Use $\boxed{\text{2nd}}$ $\boxed{\text{DRAW}}$ $\boxed{6}$ $\boxed{\text{ALPHA}}$ $\boxed{W}$ $\boxed{+}$ $\boxed{\text{ALPHA}}$ $\boxed{M}$ $\boxed{(}$ $\boxed{X|T}$ $\boxed{-}$ $\boxed{\text{ALPHA}}$ $\boxed{Z}$ $\boxed{)}$ $\boxed{\text{ENTER}}$. This command draws the tangent line $y = W + M(x - Z)$ at $x = Z$ on the graph.

22. Press $\boxed{\text{PRGM}}$ $\boxed{3}$ $\boxed{\text{ALPHA}}$ $\boxed{M}$ $\boxed{\text{2nd}}$ $\boxed{\text{TEST}}$ $\boxed{1}$ $\boxed{\emptyset}$ $\boxed{\text{ENTER}}$.
If the tangent line is horizontal and does not intersect the x-axis, then the condition $M = 0$ is true, and the next line of the program is used. The execution goes to Label 2 on line 27. A message is displayed indicating that Newton's method does not work in this case and the program stops. If the tangent line is not horizontal, the condition $M = 0$ is false, line 23 is skipped, and the program continues.

23. Press $\boxed{\text{PRGM}}$ $\boxed{2}$ $\boxed{2}$ $\boxed{\mathit{ENTER}}$.

24. Press $\boxed{\text{ALPHA}}$ $\boxed{Z}$ $\boxed{-}$ $\boxed{\text{ALPHA}}$ $\boxed{W}$ $\boxed{\div}$ $\boxed{\text{ALPHA}}$ $\boxed{M}$ $\boxed{\text{STO}\blacktriangleright}$ $\boxed{Z}$ $\boxed{\text{ENTER}}$. The x-intercept of the tangent line is stored as a new value of Z.

25. Press $\boxed{\text{ALPHA}}$ $\boxed{J}$ $\boxed{+}$ $\boxed{1}$ $\boxed{\text{STO}\blacktriangleright}$ $\boxed{J}$ $\boxed{\text{ENTER}}$. The value of J is increased by 1.

26. Press $\boxed{\text{PRGM}}$ $\boxed{2}$ $\boxed{1}$ $\boxed{\text{ENTER}}$. The execution of the program goes to Lbl 1 on line 10 and the next iteration is carried out. This is an infinite loop. The user has to interrupt it by pressing $\boxed{\text{ON}}$ $\boxed{2}$ to abort the program.

27. Press $\boxed{\text{PRGM}}$ $\boxed{1}$ $\boxed{2}$ $\boxed{\text{ENTER}}$.

28. Press $\boxed{\text{PRGM}}$ $\boxed{6}$ $\boxed{\text{ENTER}}$.

29. Press $\boxed{\text{PRGM}}$ $\boxed{\blacktriangleright}$ $\boxed{1}$ $\boxed{\text{2nd}}$ $\boxed{\text{A-LOCK}}$ $\boxed{''}$ $\boxed{S}$ $\boxed{L}$ $\boxed{O}$ $\boxed{P}$ $\boxed{E}$ $\boxed{\text{2nd}}$ $\boxed{\text{TEST}}$ $\boxed{1}$ $\boxed{\emptyset}$ $\boxed{\text{ALPHA}}$ $\boxed{''}$ $\boxed{\text{ENTER}}$.

Graphing calculator workbook   119

30. Press PRGM ▶ 1 2nd A-LOCK " T H E -
    M E T H O D - F A I L S "
    ENTER . This lines indicate that a horizontal tangent line has occurred.

Using the program

Example 1   Figure 1 shows the graph of $F(x) = \frac{1}{5}x^5 + \frac{1}{3}x^3 + x - 4$. Because $F(x)$ s continuous for all X; $F(x) \to -\infty$ as $x \to -\infty$; $F(x) \to \infty$ as $x \to \infty$; and $F'(x) = x^4 + x^2 + 1$ is positive for all x, the equation $F(x) = 0$ has exactly one solution. Find its value to the accuracy of your calculator or computer by using Newton's method with $x_0 = 1$.

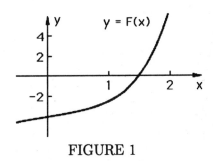
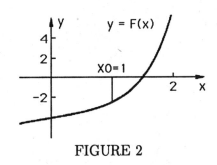

FIGURE 1   FIGURE 2

Solution   Press Y = . Enter the function $F(x)$ as $Y_1 = X \wedge 5/5 + X \wedge 3/3 + X - 4$ and its derivative as $Y_2 = X \wedge 4 + X^2 + 1$. Press RANGE and set x-min $= -0.5$, x-max $= 2.5$, x-scale $= 1$, y-min $= -6$, y-max $= 6$, and y-scale $= 2$.

Press PRGM , the number of your Newon's method program, and ENTER . When the comments and the prompt $X0 = ?$ appear, press 1   ENTER . The graph of $F(x)$ and the vertical line between it and the x-axis will be drawn as in Figure 2.

(To interrupt the program at any time , press ON and 2 (QUIT). You can then restart the program with ENTER , provided you do not press any other keys first.)

Press ENTER . The values $J = 0, X0 = 1$, and $F(X0) \doteq -2.466666667$ will be displayed. Press ENTER again. The tangent line at $x = 1$ will be drawn with a vertical line from its x-intercept to the graph as in Figure 3. Press ENTER to obtain the value $J = 1$, the value $X1 = 1.822222222$ of the x-intercept, and the value $F(X1) \doteq 3.85736232$ of the function at that point.

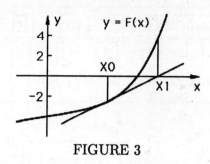
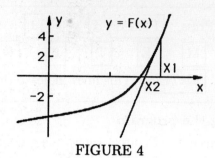

FIGURE 3   FIGURE 4

Press ENTER to see the tangent line at $x = X1$ on the curve as in Figure 4. Continue until seven tangent lines have been drawn giving the seven values of XJ and $F(XJ)$ in the table below. The picture on your calculator becomes fairly cluttered because all of the tangent lines are shown, and the last few are indistinguishable because they are so close to each other.

Notice that the values of $X6$ and $X7$ are the same. At this point the calculator sees the term $F(x_j)/F'(x_j)$ in the Newton's method formula (1) as zero, so the numbers do not change and no further accuracy is obtained by proceeding. We conclude that the approximate decimal value of the solution of $F(x) = 0$ is $x \doteq 1.482402577$ and stop the program by pressing ON 2 .

J	XJ	F(XJ)
0	1	-2.466666667
1	1.822222222	3.857362320
2	1.570865594	0.7760094784
3	1.489665542	0.0587206085
4	1.482454829	$4.19427241 \times 10^{-4}$
5	1.482402580	2.1833×10^{-8}
6	1.482402577	1×10^{-12}
7	1.482402577	1×10^{-12}

If an error message appears, press ENTER to go to the line of the program with the error so you can correct it. If the program runs but does not do what you expect, stop it with ON 2 and compare what you entered with the listing on the first two pages of this section.

Try $F(x) = x^3 - 3x^2 + 5$ with $-0.5 \leq x \leq 3, -0.5 \leq y \leq 3$ and $x_0 = 2$ for a case of a horizontal tangent line. If Newton's method takes more than six or eight steps to give an answer, the formula for the derivative $F' = Y_2$ might be wrong. The method can also be slow, even when it is used properly, if the derivative happens to be zero at the zero of the function $F(x)$. To see this, try $F(x) = (x-1)^2$ with $-0.5 \leq x \leq 3, -0.5 \leq y \leq 2$ and $x_0 = 2$.

Graphing calculator workbook

Worksheet 2J.1†

Newton's method: Finding the dimensions of a cone

Name _____ Date _____

Others in your group _____

Instructor, Teaching Assistant, and/or Recitation section _____

Directions Put first drafts of your calculations and answers on scratch paper. Take your time, work carefully, and discuss your solution with at least one other student before putting a final draft on this sheet or on other paper. Turn in all your work.

Problem 1 You want to construct a right circular cone as in Figure 1 so that it is one meter high and the total surface area of its base and lateral surface is 10 square meters.

a. If the radius of its base is r meters, then the lateral surface area is $\pi r\sqrt{1+r^2}$. Give the total area A of the base and lateral surface as a function of r. Generate the graph of this function and the line $A = 10$ on your calculator or computer, using x for r, y for A, $-0.5 \leq x \leq 3$, $-5 \leq y \leq 25$, and y-scale $= 5$. Copy the curve and line in Figure 2.

b. Use the trace command on a calculator without zooming in, or the graph on a computer to find the approximate radius for which the total surface area is 10.

c. Set $F(r) = A(r) - 10$, and use the product and chain rules to find dF/dr.

d. Use the Newton's method program 2I, with x in place of r, to find a more accurate value of the solution of $F(r) = 0$. Use $-0.5 \leq x \leq 3$, $-15 \leq y \leq 35$ for the graphs and $x_0 = 2$. In the first step of the program the graph of $F(x)$ should be generated with a vertical line from the x-axis up to the curve at $x = 2$. Then the values $J = 0$, $XJ = 2$, and $F(XJ) = 16.61600008$ will be given because you are at the zero step of Newton's method, $x_0 = 2$, and $F(x_0) \doteq 16.61600008$.

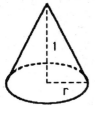

FIGURE 1

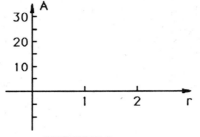

FIGURE 2

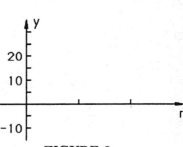

FIGURE 3

---

†This worksheet is an application of Newton's method. It uses Program 2I. See Calculator instructions 1B.3 for tips on working it with a calculator.

At the next step the tangent line at $x = 2$ will appear. Its x-intercept is x_1. Copy the curve and the tangent line in Figure 3.

Next, values x_1 and $F(x_1)$ are given. Put them in the second row of the table below. Then the tangent line at $x = x_1$ will be drawn, followed by the values of x_2 and $F(x_2)$. Put these in the table. Continue until two successive values of x_j are equal, so that continuing the procedure does not improve the answer.

e. Use the result of part (d) to solve the problem in the first paragraph.

j	$x_j \doteq$	$F(x_j) = A(x_j) - 10 \doteq$
0	2	16.61600008
1		
2		
3		
4		
5		
6		
7		
8		

Graphing calculator workbook

Worksheet 2J.2†

Newton's method: Mimimizing the light from two lamps

Name _____ Date _____

Others in your group _____

Instructor, Teaching Assistant, and/or Recitation section _____

Directions Put first drafts of your calculations and answers on scratch paper. Take your time, work carefully, and discuss your solution with at least one other student before putting a final draft on this sheet or on other paper. Turn in all your work.

Problem 1 The point P in Figure 1 is three feet above the floor in a vertical plane with two lamps that are shining on it from opposite sides. The first lamp is on the wall, eight feet above the floor, and is ten times as bright as the second lamp, which is three feet above the floor. The intensity of the light from the first lamp at P is $\dfrac{200}{D_1^2}$ candles per square foot and the intensity, at P of the light from the seond lamp is $\dfrac{20}{D_2^2}$ candles per square foot, where D_1 is the distance from P to the first lamp and D_2 is the distance from P to the second lamp.

a. Express the total intensity from both lamps as a function $I(x)$ of x. Generate its graph on your calculator or computer with y in place of I and copy it in Figure 2. Use $-1 \leq x \leq 10$, x-scale $= 2$, $-2 \leq y \leq 15$, and y-scale $= 5$.

b. What is the limit of $I(x)$ as $x \to 10^-$? Explain.

c. Use the trace command on your calculator, without zooming in, or the graph on your computer to find the approximate minimum total intensity from both lamps for x between 0 and 10 and the approximate value of x where it occurs.

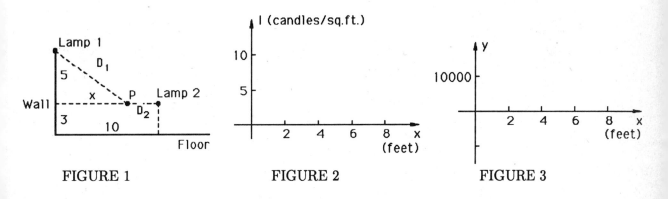

FIGURE 1   FIGURE 2   FIGURE 3

---

†This worksheet is an application of Newton's method. It uses Program 2I. See Calculator instructions 1B.3 for tips on working it with a calculator.

d. Find dI/dx:

e. Take the common denominator in the last formula to find a function $F(x)$ such that

$$\frac{dI}{dx} = \frac{40F(x)}{(10-x)^3(25+x^2)^2}.$$

f. Generate the graph of $F(x)$ and copy it in Figure 3. Use $-1 \le x \le 10$, x-scale $= 2$, $-15000 \le y \le 20000$, and y-scale $= 10000$.

g. Solve $F(x) = 0$ for $0 < x < 10$ by Newton's method using program 2I. Why does the result give the location of P for which the total intensity from the two lights is a minimum? What is that minimum?

Graphing calculator workbook

Worksheet 2J.3[†]

Newton's method: Studying the graph of a function

Name _____ Date _____

Others in your group _____

Instructor, Teaching Assistant, and/or Recitation section _____

Directions   Put first drafts of your calculations and answers on scratch paper. Take your time, work carefully, and discuss your solution with at least one other student before putting a final draft on this sheet or on other paper. Turn in all your work.

Problem 1   Generate the graph of $y(x) = 3 + 2\sin x - x^3$ on your calculator or computer for $-2 \le x \le 2$, $-4 \le y \le 8$ and copy it in Figure 1.

    a.   Use the trace and zoom commands on a calculator or the graph on a computer to find the approximate maximum value of $y(x)$ for $x \ge 0$, the approximate value of x where the maximum occurs, and the approximate coordinates of the inflection point of its graph that is closest to the y-axis.

    b.   Generate the graph of $\dfrac{dy}{dx}$ and copy it in Figure 2. Use $-2 \le x \le 2, -4 \le y \le 3$.

    c.   Generate the graph of $\dfrac{d^2y}{dx^2}$ and copy it in Figure 3. Use $-2 \le x \le 2, -15 \le y \le 15$.

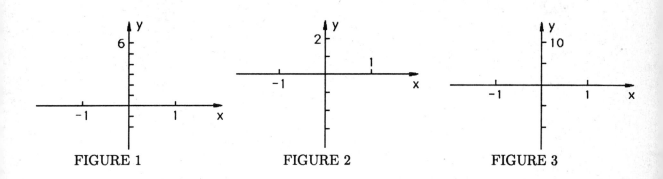

           FIGURE 1   FIGURE 2   FIGURE 3

[†]This worksheet is an application of Newton's method. It uses Program 2I. See Calculator instructions 1B.3 for tips on working it with a calculator.

d. Use Newton's method, as needed, to find the numbers in part (a) with as much accuracy as your calculator or computer provides.

Graphing calculator workbook

Worksheet 2J.4[†]

Newton's method: Finding the closest point to a curve

Name _____ Date _____

Others in your group _____

Instructor, Teaching Assistant, and/or Recitation section _____

Directions Put first drafts of your calculations and answers on scratch paper. Take your time, work carefully, and discuss your solution with at least one other student before putting a final draft on this sheet or on other paper. Turn in all your work.

Problem 1 Pick a function $g(x)$ that is differentiable and has positive values for $x > 0$ and whose graph is not a part of a circle or line. Generate its graph on your calculator or computer for $0 \leq x \leq x$-max, $0 \leq y \leq y$-max with x-max equal to one-and-a-half times y-max, so the scales are equal on the axes. Adjust the formula for the function and your choices of x-max and y-max until you like the appearance of your graph. Copy it in Figure 1 and put scales on the axes.

Next, pick a point $P = (a, b)$ on your drawing that is far enough from the graph to look good but close enough so that the closest point on the graph to P will also be in the drawing. Plot the point P in Figure 1.

a. Express the distance from P to $(x, g(x))$ as a function $D(x)$.

b. Generate the graph of $D(x)$ with the range of x from Figure 1 and with a suitable range of D. Copy it in Figure 2 and put scales on the axes.

c. Give a formula for $S(x) = D(x)^2$.

d. Generate the graph of $S(x)$ with the range of x from Figure 1 and a suitable range for S. Copy it in Figure 3 and put scales on the axes.

e. Why does the minimum of $S(x)$ occur at the same value of x as the minimum of $D(x)$?

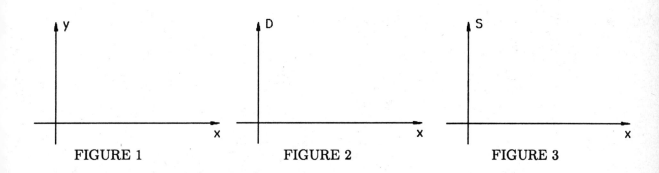

FIGURE 1   FIGURE 2   FIGURE 3

[†]This worksheet uses the Newton's method program 2I. See Calculator instructions 1B.3 for tips on working it with a calculator.

f. Find a formula for $F(x) = \dfrac{dS}{dx}$

g. Generate the graph of $\dfrac{dS}{dx}$ with the same x-range as in Figures 1 through 3 and with a suitable y-range. Copy it in Figure 4 and put scales on the axes.

h. Apply Newton's method to find, with as much accuracy as your calculator or computer provides, the coordinates of the closest point to P on the graph of $g(x)$ and the distance from that point to P.

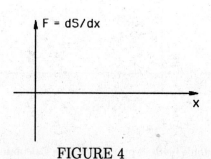

FIGURE 4

Graphing calculator workbook

Worksheet 3A.1†

Finding distances from velocities: Initial investigations

Name _____ Date _____

Others in your group _____

Instructor, Teaching Assistant, and/or Recitation section _____

Directions Put first drafts of your calculations and answers on scratch paper. Take your time, work carefully, and discuss your solution with at least one other student before putting a final draft on this sheet or on other paper. Turn in all your work.

Problem 1 You drive east on Highway 200 from Turtle Lake, North Dakota, traveling 20 miles per hour for two hours, 60 miles per hour during the third hour, and 40 miles per hour for three more hours. Figure 1 shows the graph of your velocity toward the east as a function of time (hours) with $t = 0$ at the beginning of your trip.

    a.   How far do you travel in the first two hours, from $t = 0$ to $t = 2$?

    b.   How far do you travel in the third hour, from $t = 2$ to $t = 3$?

    c.   How far do you travel in the last three hours, from $t = 3$ to $t = 6$?

    d.   How far do you travel in the six hours from $t = 0$ to $t = 6$?

    e.   How is the answer to part (d) related to the areas of the rectangles in Figure 2?

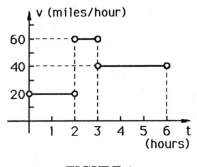 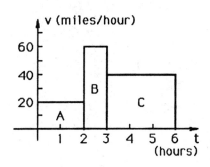

                  FIGURE 1   FIGURE 2

Problem 2 The next day, you continue east on Highway 200 from Zerkel, Minnesota, driving 40 miles per hour for two hours. Then you decide to return to Mahnomen, Minnesota and drive west for two hours at 20 miles per hour and two hours at 30 miles per hour. The graph of your velocity toward the east is shown in Figure 3.

    a.   How far do you travel toward the east from $t = 0$ to $t = 2$?

    b.   How far do you travel toward the east from $t = 2$ to $t = 4$? (The answer is negative because you actually travel toward the west.)

---

†These problems use the formula, [Distance] = [Time][Velocity], for motion under constant velocity to solve in special cases a type of problem that usually requires integrals.

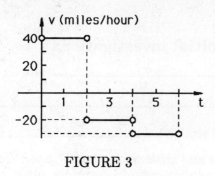

FIGURE 3

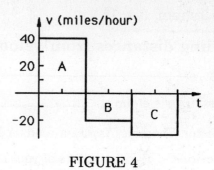

FIGURE 4

- **c.** How far do you travel toward the east from $t = 4$ to $t = 6$?
- **d.** How many miles east are you at $t = 6$ from where you were at $t = 0$?
- **e.** What is the total distance you travel in the six hours?
- **f.** How are the answers to parts (d) and (e) related to the areas of the rectangles in Figure 3?

Problem 3 Another day you begin by driving east on Highway 82 from Paris, Texas. Figure 5 shows the graph of your velocity toward the east with $t = 0$ at the beginning of your day's drive.

- **a.** How far do you travel toward the east from $t = 0$ to $t = 2$?
- **b.** How far do you travel toward the east from $t = 2$ to $t = 4$?
- **c.** How far do you travel toward the east from $t = 4$ to $t = 6$?
- **d.** How many miles east are you at $t = 6$ from where you were at $t = 0$?
- **e.** What is the total distance you travel in the six hours?
- **f.** How are the answers to parts (d) and (e) related to the areas of the rectangles in Figure 6?

The velocity functions in these examples are called "piecewise constant" or "step" functions. Notice that the tops and bottoms of the rectangles in Figures 2, 4, and 6 are formed by the t-axis and the graphs of the corresponding functions in Figures 1, 3, and 5. Also, notice that in each case the net distance traveled equals the sum of the areas of the rectangles above the t-axis minus the areas of the rectangles below the t-axis, and the total distance traveled is the sum of the areas of all the rectangles. You will learn that this relationship between areas and distance generalizes to more general types of velocity functions through the use of integrals.

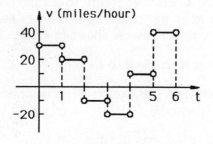

FIGURE 5

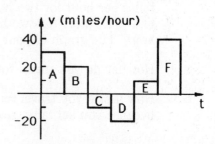

FIGURE 6

Graphing calculator workbook   131

Worksheet 3A.2†

Volumes by slicing: Initial investigations

Name _____ Date _____

Others in your group _____

Instructor, Teaching Assistant, and/or Recitation section _____

Directions   Put first drafts of your calculations and answers on scratch paper. Take your time, work carefully, and discuss your solution with at least one other student before putting a final draft on this sheet or on other paper. Turn in all your work.

Problem 1   A jacuzzi has the shape of a right circular cylinder of radius 5 feet and height 2 feet on top of a right circular cylinder of radius 3 feet and height 4 feet (Figure 1).

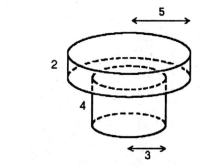

FIGURE 1

a.   What is the total volume of the jacuzzi?

b.   Figure 2 shows the area $A(h)$ of horizontal cross sections of the jacuzzi as a function of the height h above the bottom. Because the cross section is a circle of radius 3 for $0 < h < 4$ and a circle of radius 5 for $4 < h < 6$, $A(h) = 9\pi$ for $0 < h < 4$ and $A(h) = 25\pi$ for $4 < h < 6$. How is the result of part (a) related to the areas of the two rectangles in Figure 3?

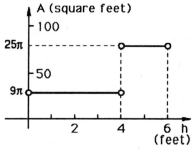

FIGURE 2

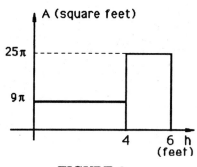

FIGURE 3

---

†These problems use formulas for volumes of cylinders and boxes to solve in special cases a type of problem that usually requires integrals.

Problem 2 An office building has the shape of three rectangular boxes (Figure 4). The section on the south is 30 feet wide, 30 feet high, and 60 feet long. The middle section is 60 feet wide, 60 feet high, and 40 feet long. The section on the north is 80 feet wide, 80 feet high, and 60 feet long.

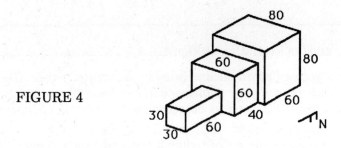

FIGURE 4

a. What is the total volume of the building?

b. Figure 5 shows the area $A(x)$ of east-west vertical cross sections of the building as a function of the distance x from the south end. Because the cross section is a square of width 30 for $0 < x < 60$, a square of width 60 for $60 < x < 100$, and a square of width 80 for $100 < x < 160$, $A(x) = 900$ for $0 < h < 60$, $A(x) = 3600$ for $60 < x < 100$, and $A(x) = 6400$ for $100 < x < 160$. How is the result of part (a) related to the areas of the three rectangles in Figure 6?

The functions $A(x)$ giving the areas of the cross sections in these examples are piecewise constant (step) functions. Notice that the tops of the rectangles in Figures 3 and 6 are formed by the graphs of the functions in Figures 2 and 5, and that in each case the volume equals the sum of the areas of the rectangles. You will see that this relationship between volumes and cross sectional areas generalizes to more general types of solids through the use of integrals.

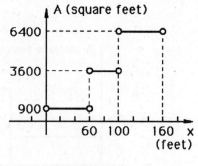

FIGURE 5

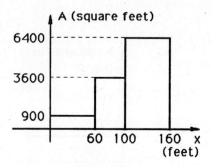

FIGURE 6

Graphing calculator workbook

Worksheet 3A.3[†]

Lengths of curves and average values: Initial investigations

Name _____ Date _____

Others in your group _____

Instructor, Teaching Assistant, and/or Recitation section _____

Directions *Put first drafts of your calculations and answers on scratch paper. Take your time, work carefully, and discuss your solution with at least one other student before putting a final draft on this sheet or on other paper. Turn in all your work.*

Problem 1 The graph of the function $f(x)$ in Figure 1 consists of three line segments for $0 \le x \le 8$.

    a.   What is the total length of the graph?

    b.   What is $f'(x)$ for $0 < x < 3$?

    c.   What is $f'(x)$ for $3 < x < 4$?

    d.   What is $f'(x)$ for $4 < x < 8$?

    e.   Figure 2 shows the graph of $\sqrt{1 + [f'(x)]^2}$. What are the y-coordinates of the three lines that form its graph?

    f.   Find the areas of the three rectangles in Figure 3.

    g.   How are the answers to parts (a) and (f) related?

The function in Figure 1 is called "piecewise linear." You will learn that the result of part (g) generalizes to other types of functions through the use of integrals.

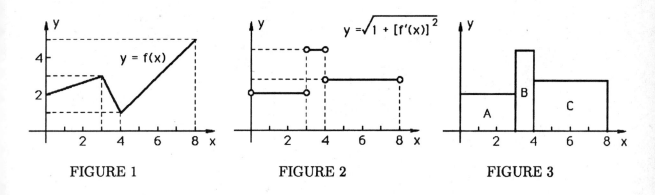

FIGURE 1   FIGURE 2   FIGURE 3

[†]This worksheeet uses the Pythagorean theorem and the formula for the area of a rectangle to solve in special cases problems that usually require integrals.

Problem 2   Figure 4 shows the graph of the maximum daily temperature $T(t)$ in a northern city during a week in the winter, given as a function of the time t measured in days with $t = 0$ at 12:00 midnight Sunday. The maximum temperature was 20° Fahrenheit each day on Sunday and Monday (i.e., for $0 < t < 2$) and 10° on Tuesday (i.e., for $2 < t < 3$). There was a cold spell Wednesday, when the maximum temperature was −20°, and it warmed up on Friday and Saturday to a maximum temperature of 30° each day.

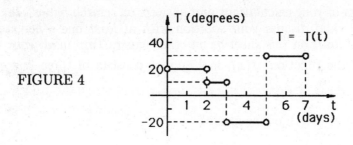

FIGURE 4

a.   What was the average maximum daily temperature during the week?

b.   Find the areas of the rectangles in Figure 5.

c.   How are the areas of the rectangles in Figure 5 related to the answer to part (a)?

You will learn that the result of part (c) generalizes to other types of functions with the use of integrals.

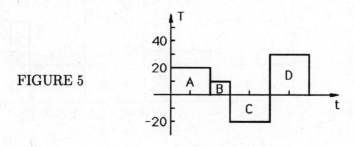

FIGURE 5

Riemann sum program 3B

Texas Instruments TI-81

Because the function $f(x)$ of Figure 1 is ≥ 0 for $a \leq x \leq c$ and is ≤ 0 for $c \leq x \leq b$, its definite (Riemann) integral $\int_a^b f(x)\,dx$ from $x = a$ to $x = b$ equals the area of region A above the x-axis in the figure minus the area of region B below the x-axis.

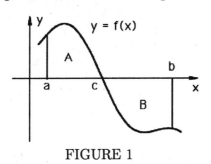 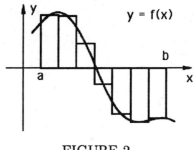

FIGURE 1   FIGURE 2

A Riemann sum approximation of the integral is obtained by approximating the regions by vertical rectangles that have one end on the x-axis and touch the graph of the function at the other end. Figure 2 shows an approximation by seven rectangles of equal widths that touch the graph at the midpoints of their ends. The Riemann sum equals the sum of the areas of the three rectangles above the x-axis minus the areas of the four rectangles below the x-axis. The integral is the limit of such Riemann sums as the number of rectangles tends to infinity and their widths tend to 0.

The following program draws the graph and rectangles and calculates Riemann sums for the function given as Y_1 in the Y= menu. The user gives $a = A$, $b = B$, the number N of subintervals, and a number T between 0 and 1. The rectangles are of equal widths and touch the graph at their upper or lower left corners if $T = 0$, at their upper or lower right corners if $T = 1$, and at the midpoints of their tops or bottoms if $T = 0.5$.

The program (214 bytes)

1.   Prgm 3: RIEMANN
2.   : ClrDraw
3.   : All-Off
4.   : Y_1-On
5.   : Disp "USE $Y_1(X)$"
6.   : Disp "$A =$"
7.   : Input A
8.   : Disp "$B =$"
9.   : Input B
10.   : Disp "$N =$"
11.   : Input N
12.   : Disp "$T = \emptyset$ FOR LEFT"   (Note: $\emptyset$ denotes zero.)
13.   : Disp "$T = 1$ FOR RIGHT"
14.   : Disp "$T = \emptyset.5$ FOR MID"
15.   : Input T

16. $: (B - A)/N \to D$ (Note: Use $\boxed{-}$ not $\boxed{(-)}$.)

17. $: \emptyset \to S$

18. $: 1 \to J$

19.   : Lbl 1

20. $: A + (J - 1)D \to U$

21. $: U + TD \to X$

22. $: Y_1 \to W$

23. $: \text{Line}(U, \emptyset, U, W)$

24. $: \text{Line}(U, W, U + D, W)$

25. $: \text{Line}(U + D, W, U + D, \emptyset)$

26. $: S + DW \to S$

27. $: J + 1 \to J$

28. $: \text{If } J \leq N$

29.   : Goto 1

30.   : Pause

31.   : Disp "SUM ="

32.   : Disp S

33. $: S + 0$

Entering the program

Press $\boxed{\text{2nd}}$ $\boxed{\text{QUIT}}$ to display the home screen and $\boxed{\text{PRGM}}$ for the menu of programs. Press $\boxed{\blacktriangleright}$ to highlight EDIT and $\boxed{3}$ for program #3. Instructions for entering the program follow. Use $\boxed{\text{INS}}$ and $\boxed{\text{DEL}}$ to make corrections. To return to the program from other screens, press $\boxed{\text{CLEAR}}$ or follow the steps above.

---

1.   Press $\boxed{\text{R}}$ $\boxed{\text{I}}$ $\boxed{\text{E}}$ $\boxed{\text{M}}$ $\boxed{\text{A}}$ $\boxed{\text{N}}$ $\boxed{\text{N}}$ $\boxed{\text{ENTER}}$. The calculator is automatically in alpha mode here. This title identifies the program.

---

2.   Press $\boxed{\text{2nd}}$ $\boxed{\text{DRAW}}$ $\boxed{1}$ $\boxed{\text{ENTER}}$. ClrDraw clears the graphics screen.

---

3.   Press $\boxed{\text{2nd}}$ $\boxed{\text{Y-Vars}}$ $\boxed{\blacktriangleleft}$ $\boxed{1}$ $\boxed{\text{ENTER}}$ to unselect the functions in the $Y=$ menu.

---

4.   Press $\boxed{\text{2nd}}$ $\boxed{\text{Y-VARS}}$ $\boxed{\blacktriangleright}$ $\boxed{2}$ $\boxed{\text{ENTER}}$ to select Y_1.

Graphing calculator workbook 137

5. Press $\boxed{\text{PRGM}}$ $\boxed{\blacktriangleright}$ $\boxed{1}$ for the word Disp and then $\boxed{\text{2nd}}$ $\boxed{\text{A-LOCK}}$ $\boxed{\text{``}}$
$\boxed{\text{U}}$ $\boxed{\text{S}}$ $\boxed{\text{E}}$ $\boxed{\text{-}}$ $\boxed{\text{2nd}}$ $\boxed{\text{Y-VARS}}$ $\boxed{1}$ $\boxed{(}$ $\boxed{X|T}$ $\boxed{)}$ $\boxed{\text{ALPHA}}$
$\boxed{\text{``}}$ $\boxed{\text{ENTER}}$. The space is on the key with 0.

6. Press $\boxed{\text{PRGM}}$ $\boxed{\blacktriangleright}$ $\boxed{1}$ $\boxed{\text{ALPHA}}$ $\boxed{\text{``}}$ $\boxed{\text{ALPHA}}$ $\boxed{A}$ $\boxed{\text{2nd}}$ $\boxed{\text{TEST}}$
$\boxed{1}$ $\boxed{\text{ALPHA}}$ $\boxed{\text{``}}$ $\boxed{\text{ENTER}}$ to cause the prompt $A = ?$ to appear when the program is run.

7. Press $\boxed{\text{PRGM}}$ $\boxed{\blacktriangleright}$ $\boxed{2}$ $\boxed{\text{ALPHA}}$ $\boxed{A}$ $\boxed{\text{ENTER}}$. Here the program pauses for the user to assign a value to A.

8. Press $\boxed{\text{PRGM}}$ $\boxed{\blacktriangleright}$ $\boxed{1}$ $\boxed{\text{ALPHA}}$ $\boxed{\text{``}}$ $\boxed{\text{ALPHA}}$ $\boxed{B}$ $\boxed{\text{2nd}}$ $\boxed{\text{TEST}}$
$\boxed{1}$ $\boxed{\text{ALPHA}}$ $\boxed{\text{``}}$ $\boxed{\text{ENTER}}$. The prompt $B = ?$ will appear.

9. Press $\boxed{\text{PRGM}}$ $\boxed{\blacktriangleright}$ $\boxed{2}$ $\boxed{\text{ALPHA}}$ $\boxed{B}$ $\boxed{\text{ENTER}}$. The user gives B.

10. Press $\boxed{\text{PRGM}}$ $\boxed{\blacktriangleright}$ $\boxed{1}$ $\boxed{\text{ALPHA}}$ $\boxed{\text{``}}$ $\boxed{\text{ALPHA}}$ $\boxed{N}$ $\boxed{\text{2nd}}$ $\boxed{\text{TEST}}$
$\boxed{1}$ $\boxed{\text{ALPHA}}$ $\boxed{\text{``}}$ $\boxed{\text{ENTER}}$. The prompt $N = ?$ will appear.

11. Press $\boxed{\text{PRGM}}$ $\boxed{\blacktriangleright}$ $\boxed{2}$ $\boxed{\text{ALPHA}}$ $\boxed{N}$ $\boxed{\text{ENTER}}$. The user chooses N.

12. Press $\boxed{\text{PRGM}}$ $\boxed{\blacktriangleright}$ $\boxed{1}$ $\boxed{\text{ALPHA}}$ $\boxed{\text{``}}$ $\boxed{\text{ALPHA}}$ $\boxed{T}$ $\boxed{\text{2nd}}$ $\boxed{\text{TEST}}$
$\boxed{1}$ $\boxed{\emptyset}$ $\boxed{\text{2nd}}$ $\boxed{\text{A-LOCK}}$ $\boxed{\text{-}}$ $\boxed{F}$ $\boxed{O}$ $\boxed{R}$ $\boxed{\text{-}}$ $\boxed{L}$ $\boxed{E}$ $\boxed{F}$
$\boxed{T}$ $\boxed{\text{``}}$ $\boxed{\text{ENTER}}$. This and the next two lines explain the role of T.

13. Press $\boxed{\text{PRGM}}$ $\boxed{\blacktriangleright}$ $\boxed{1}$ $\boxed{\text{ALPHA}}$ $\boxed{\text{``}}$ $\boxed{\text{ALPHA}}$ $\boxed{T}$ $\boxed{\text{2nd}}$ $\boxed{\text{TEST}}$
$\boxed{1}$ $\boxed{1}$ $\boxed{\text{2nd}}$ $\boxed{\text{A-LOCK}}$ $\boxed{\text{-}}$ $\boxed{F}$ $\boxed{O}$ $\boxed{R}$ $\boxed{\text{-}}$ $\boxed{R}$ $\boxed{I}$ $\boxed{G}$
$\boxed{H}$ $\boxed{T}$ $\boxed{\text{``}}$ $\boxed{\text{ENTER}}$.

14. Press $\boxed{\text{PRGM}}$ $\boxed{\blacktriangleright}$ $\boxed{1}$ $\boxed{\text{ALPHA}}$ $\boxed{\text{``}}$ $\boxed{\text{ALPHA}}$ $\boxed{T}$ $\boxed{\text{2nd}}$ $\boxed{\text{TEST}}$
$\boxed{1}$ $\boxed{\emptyset}$ $\boxed{.}$ $\boxed{5}$ $\boxed{\text{2nd}}$ $\boxed{\text{A-LOCK}}$ $\boxed{\text{-}}$ $\boxed{F}$ $\boxed{O}$ $\boxed{R}$ $\boxed{\text{-}}$ $\boxed{M}$
$\boxed{I}$ $\boxed{D}$ $\boxed{\text{``}}$ $\boxed{\text{ENTER}}$.

15. Press $\boxed{\text{PRGM}}$ $\boxed{\blacktriangleright}$ $\boxed{2}$ $\boxed{\text{ALPHA}}$ $\boxed{T}$ $\boxed{\text{ENTER}}$. The user picks T.

16. Press $\boxed{(}$ $\boxed{\text{ALPHA}}$ $\boxed{B}$ $\boxed{-}$ $\boxed{\text{ALPHA}}$ $\boxed{A}$ $\boxed{)}$ $\boxed{\div}$ $\boxed{\text{ALPHA}}$ $\boxed{N}$ $\boxed{\text{STO}\blacktriangleright}$ $\boxed{D}$ $\boxed{\text{ENTER}}$. The width D of the rectangles is calculated and stored.

17. Press $\boxed{\emptyset}$ $\boxed{\text{STO}\blacktriangleright}$ $\boxed{S}$ $\boxed{\text{ENTER}}$. The Riemann sum is stored as S during the calculations. Here S is set equal to 0 to erase its previous value.

18. Press $\boxed{1}$ $\boxed{\text{STO}\blacktriangleright}$ $\boxed{J}$ $\boxed{\text{ENTER}}$. Set $J = 1$ for the first rectangle.

19. Press $\boxed{\text{PRGM}}$ $\boxed{1}$ $\boxed{1}$ $\boxed{\text{ENTER}}$. This is the beginning of a loop that ends with the Goto statement on line 29.

20. Press $\boxed{\text{ALPHA}}$ $\boxed{A}$ $\boxed{+}$ $\boxed{(}$ $\boxed{\text{ALPHA}}$ $\boxed{J}$ $\boxed{-}$ $\boxed{1}$ $\boxed{)}$ $\boxed{\text{ALPHA}}$ $\boxed{D}$ $\boxed{\text{STO}\blacktriangleright}$ $\boxed{U}$ $\boxed{\text{ENTER}}$. U is the x-coordinate of the left side of the rectangle.

21. Press $\boxed{\text{ALPHA}}$ $\boxed{U}$ $\boxed{+}$ $\boxed{\text{ALPHA}}$ $\boxed{T}$ $\boxed{\text{ALPHA}}$ $\boxed{D}$ $\boxed{\text{STO}\blacktriangleright}$ $\boxed{X|T}$ $\boxed{\text{ENTER}}$. Here X is the x-coordinate of the left edge of the rectangle if $T = 0$, the right edge if $T = 1$, and the midpoint if $T = 0.5$.

22. Press $\boxed{\text{2nd}}$ $\boxed{\text{Y-vars}}$ $\boxed{1}$ $\boxed{\text{STO}\blacktriangleright}$ $\boxed{W}$ $\boxed{\text{ENTER}}$. The value at X of the function Y_1 in the Y= menu is calculated and stored as W.

23. Press $\boxed{\text{2nd}}$ $\boxed{\text{DRAW}}$ $\boxed{2}$ $\boxed{\text{ALPHA}}$ $\boxed{U}$ $\boxed{\text{ALPHA}}$ $\boxed{,}$ $\boxed{\emptyset}$ $\boxed{\text{ALPHA}}$ $\boxed{,}$ $\boxed{\text{ALPHA}}$ $\boxed{U}$ $\boxed{\text{ALPHA}}$ $\boxed{,}$ $\boxed{\text{ALPHA}}$ $\boxed{W}$ $\boxed{)}$ $\boxed{\text{ENTER}}$. The left side of the rectange is drawn.

24. Press $\boxed{\text{2nd}}$ $\boxed{\text{DRAW}}$ $\boxed{2}$ $\boxed{\text{ALPHA}}$ $\boxed{U}$ $\boxed{\text{ALPHA}}$ $\boxed{,}$ $\boxed{\text{ALPHA}}$ $\boxed{W}$ $\boxed{\text{ALPHA}}$ $\boxed{,}$ $\boxed{\text{ALPHA}}$ $\boxed{U}$ $\boxed{+}$ $\boxed{\text{ALPHA}}$ $\boxed{D}$ $\boxed{\text{ALPHA}}$ $\boxed{,}$ $\boxed{\text{ALPHA}}$ $\boxed{W}$ $\boxed{)}$ $\boxed{\text{ENTER}}$. The top of the rectangle is drawn.

25. Press $\boxed{\text{2nd}}$ $\boxed{\text{DRAW}}$ $\boxed{2}$ $\boxed{\text{ALPHA}}$ $\boxed{U}$ $\boxed{+}$ $\boxed{\text{ALPHA}}$ $\boxed{D}$ $\boxed{\text{ALPHA}}$ $\boxed{,}$ $\boxed{\text{ALPHA}}$ $\boxed{W}$ $\boxed{\text{ALPHA}}$ $\boxed{,}$ $\boxed{\text{ALPHA}}$ $\boxed{U}$ $\boxed{+}$ $\boxed{\text{ALPHA}}$ $\boxed{D}$ $\boxed{\text{ALPHA}}$ $\boxed{,}$ $\boxed{\emptyset}$ $\boxed{)}$ $\boxed{\text{ENTER}}$. The right side of the rectangle is drawn.

26. Press $\boxed{\text{ALPHA}}$ $\boxed{S}$ $\boxed{+}$ $\boxed{\text{ALPHA}}$ $\boxed{D}$ $\boxed{\text{ALPHA}}$ $\boxed{W}$ $\boxed{\text{STO}\blacktriangleright}$ $\boxed{S}$ $\boxed{\text{ENTER}}$. The area of the rectangle is added to S if W is positive and is subtracted from S if W is negative.

Graphing calculator workbook   139

---

27. Press $\boxed{\text{ALPHA}}$ $\boxed{J}$ $\boxed{+}$ $\boxed{1}$ $\boxed{\text{STO}\blacktriangleright}$ $\boxed{J}$ $\boxed{\text{ENTER}}$. J is increased by 1.

---

28. Press $\boxed{\text{PRGM}}$ $\boxed{3}$ $\boxed{\text{ALPHA}}$ $\boxed{J}$ $\boxed{\text{2nd}}$ $\boxed{\text{TEST}}$ $\boxed{6}$ $\boxed{\text{ALPHA}}$ $\boxed{N}$

$\boxed{\text{ENTER}}$. If J is $\leq N$, line 29 is executed and the loop is repeated. If $J = N + 1$, the program skips to line 30.

---

29. Press $\boxed{\text{PRGM}}$ $\boxed{2}$ $\boxed{1}$ $\boxed{\text{ENTER}}$.

---

30. Press $\boxed{\text{PRGM}}$ $\boxed{6}$ $\boxed{\text{ENTER}}$ to have the program pause.

---

31. Press $\boxed{\text{PRGM}}$ $\boxed{\blacktriangleright}$ $\boxed{1}$ $\boxed{\text{2nd}}$ $\boxed{\text{A-LOCK}}$ $\boxed{``}$ $\boxed{S}$ $\boxed{U}$ $\boxed{M}$ $\boxed{\text{2nd}}$

$\boxed{\text{TEST}}$ $\boxed{1}$ $\boxed{\text{ALPHA}}$ $\boxed{``}$ $\boxed{\text{ENTER}}$. The symbols SUM = are displayed.

---

32. Press $\boxed{\text{PRGM}}$ $\boxed{\blacktriangleright}$ $\boxed{1}$ $\boxed{\text{ALPHA}}$ $\boxed{S}$ $\boxed{\text{ENTER}}$. The value of the Riemann sum is displayed.

---

33. Press $\boxed{\text{ALPHA}}$ $\boxed{S}$ $\boxed{+}$ $\boxed{0}$ $\boxed{\text{ENTER}}$. The value S of the Riemann sum is stored as *Ans* for subsequent calculations.

Using the program

Example 1   Run the program to find the following Riemann sums for $\int_0^1 (1 - x^3)\ dx$ with rectangles of equal widths. Use $-0.25 \leq x \leq 1.25, -0.25 \leq y \leq 1.25$ for the graphs and give the Riemann sums with five decimal place accuracy.

   a.   Use 6 rectangles that touch the curve at their left corners.
   b.   Use 6 rectangles that touch the curve at the their right corners.
   c.   Use 6 rectangles that touch the curve at the midpoints of their tops.
   d.   Use 20 rectangles that touch the curve at the midpoints of their tops.
   e.   Use 30 rectangles that touch the curve at the midpoints of their tops.

Solution   Press $\boxed{\text{2nd}}$ $\boxed{\text{QUIT}}$ to return to the home screen if necessary. Press

$\boxed{\text{RANGE}}$ and enter x-min $= -0.25$, x-max $= 1.25$, x-scale $= 1$, y-min $= -0.25$,

y-max $= 1.25$, and y-scale $= 1$. Press $\boxed{Y =}$, enter $1 - X \wedge 3$, and delete any extra symbols to define Y_1.

   a.   Press $\boxed{\text{PRGM}}$, $\boxed{3}$ (or the number you gave the program), and $\boxed{\text{ENTER}}$.

The prompt $A =?$ should appear. Press $\boxed{0}$ $\boxed{\text{ENTER}}$ to assign the value

0 to A, $\boxed{1}$ $\boxed{\text{ENTER}}$ to assign the value 1 to B, $\boxed{6}$ $\boxed{\text{ENTER}}$ to set

$N = 6$, and $\boxed{0}$ $\boxed{\text{ENTER}}$ to set $T = 0$. The curve and the six rectangles should be drawn as in Figure 3. Press ENTER for the value 0.82639 of the Riemann sum rounded to five decimal places.

If there is an error in the program that the calculator can locate, an error message will be given. Press 1 to go to the line with the error and correct it. If the program does not do what you expect but no error message is displayed, press ON 2 (Quit) to abort it, and compare the program you entered with the listing on the first pages of these instructions.

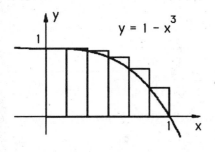

Sum ≐ 0.82639
FIGURE 3

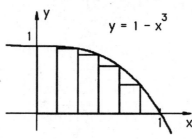

Sum ≐ 0.65972
FIGURE 4

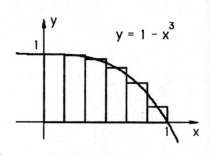

Sum ≐ 0.75347
FIGURE 5

b–e Press ENTER to run the program again. Use $A = 0, B = 1, N = 6, T = 1$ for part (b) (Figure 4), $A = 0, B = 1, N = 6, T = 0.5$ for part (c) (Figure 5), $A = 0, B = 1, N = 20, T = 0.5$ for part (d) (Figure 6), and $A = 0, B = 1, N = 30, T = 0.5$ for part (e) (Figure 7).

The exact value of the integral, which equals the area of the region in Figure 8, is the limit 0.75 of the Riemann sums as the number of rectangles used tends to ∞.

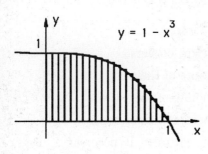

Sum ≐ 0.75031
FIGURE 6

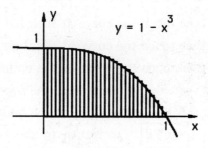

Sum ≐ 0.75014
FIGURE 7

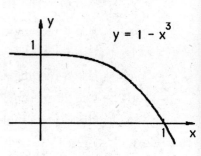

Area = 0.75
FIGURE 8

Graphing calculator workbook

Worksheet 3C.1†

Integrals of x^n: Numerical experiments

Name _____ Date _____

Others in your group _____

Instructor, Teaching Assistant, and/or Recitation section _____

Directions Put first drafts of your calculations and answers on scratch paper. Take your time, work carefully, and discuss your solution with at least one other student before putting a final draft on this sheet or on other paper. Turn in all your work.

Problem 1a Use Riemann sum program 3B with $-0.25 \leq x \leq 1.25$, $-0.25 \leq y \leq 1.25$ to calculate Riemann sums for $\int_0^1 x^2 \, dx$ corresponding to $N = 5$, $N = 10$, $N = 30$, and $N = 100$ rectangles of equal widths that touch the graph at the midpoints of their tops (the Midpoint rule). (Use $A = 0$, $B = 1$, and $T = 0.5$ each time in the program.) The Riemann sums, rounded to six decimal places, should be the numbers in the first row of the table below. Copy the curve and the rectangles for $N = 5$ in Figure 1.

    b. Pick an integer $n > 2$. Follow the instructions for part (a) with $y = x^n$ in place of $y = x^2$. Put the value of n and the Riemann sums, rounded to six decimal places, in the second row of the table. Copy the curve and the rectangles for $N = 5$ in Figure 2.

n	$N = 5$	$N = 10$	$N = 30$	$N = 100$
2	0.330000	0.332500	0.333241	0.333325

Riemann sum for $\int_0^1 x^n \, dx$ with N rectangles and $T = 0.5$

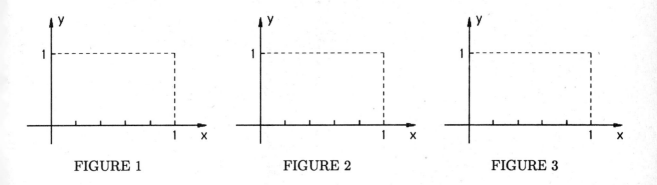

FIGURE 1   FIGURE 2   FIGURE 3

---

†This worksheet uses Riemann sum program 3B to predict values of certain integrals. See Calculator instructions 1B.3 for tips on working it with a calculator.

142   Graphing calculator workbook

c. Pick a positive fraction $n < 1$. Follow the instructions for Problem 2 with $y = x^n$ in place of $y = x^2$. Use the third row of the table and Figure 3.

d. Use the table to predict formulas for $\int_0^1 x^n \, dx$ for the three values of n and then for any $n > 0$. Calculate Riemann sums for other values of $n > 0$ if you cannot see an answer from the three examples in the table. (The answer involves $n + 1$.)

Problem 2a Use Program 3B to find the Riemann sums for $\int_0^1 x^n \, dx$ with $n = 1, 2$, and 3 using twenty subintervals ($N = 20$) and evaluating the functions at the left endpoints of the subintervals ($T = 0$). Put the values in the second column of the table below. Then find the Riemann sums for $\int_0^{10} x^n \, dx$ and $\int_0^{100} x^n \, dx$ with $N = 20, T = 0$ and $n = 1, 2$, and 3. Put their values in the third and fourth columns. You do not have to change the ranges of x and y since you only need the values of the Riemann sums and not the associated graphs.

n	Riemann sum for $\int_0^1 x^n \, dx$	Riemann sum for $\int_0^{10} x^n \, dx$	Riemann sum for $\int_0^{100} x^n \, dx$
1			
2			
3			

b. The table should show that

$$\int_0^{10} x^n \, dx = 10^{n+1} \int_0^1 x^n \, dx \quad \text{and} \quad \int_0^{100} x^n \, dx = 100^{n+1} \int_0^1 x^n \, dx \qquad (1)$$

for $n = 1, 2$, and 3. Explain equations (1) by figuring out what changing the upper limit of integration from 1 to 10 and from 1 to 100 does to the widths and heights of the twenty rectangles whose areas give the Riemann sums.

c. Use equations (1) and the result of Problem 1 to predict formulas for $\int_0^{10} x^n \, dx$ and $\int_0^{100} x^n \, dx$.

Graphing calculator workbook

Worksheet 3C.2[†]

Integrals of $\cos x$ and $\sin x$: **Numerical experiments**

Name _____ Date _____

Others in your group _____

Instructor, Teaching Assistant, and/or Recitation section _____

Directions   *Put first drafts of your calculations and answers on scratch paper. Take your time, work carefully, and discuss your solution with at least one other student before putting a final draft on this sheet or on other paper. Turn in all your work.*

Problem 1   Your goal in this problem is to predict a formula for the integral

$$\int_0^b \cos x \, dx \tag{1}$$

as a function of the upper limit of integration b, by calculating Riemann sum approximations for several of values b.

   a.   The integral (1) is 0 for $b = 0$ because its limits of integration are equal. Use the Riemann sum program 3B with 20 subintervals ($N = 20$) and with the function evaluated at the midpoints ($T = 0.5$) to find approximate values of integral (1) for $b = 0.5, 1, 1.5, 2, 2.5, 3, 3.5, 4, 4.5, 5, 5.5, 6$, and 6.5. Use $-1 \le x \le 7$, $-0.5 \le y \le 2.3$ for the graphs and $A = 0$, $B = b$, $N = 20$, and $T = 0.5$ in the program. Complete Table 1 with the values of the Riemann sums, rounded to two decimal places. Then plot the points with these coordinates in Figure 1.

   b.   The integral (1) can be expressed in terms of $\sin b$ or $\cos b$. Use Figure 1 and Table 1 to predict the formula.

      (Similar calculations could be used to predict a formula for $\displaystyle\int_0^b \sin x \, dx$.)

b	Riemann sum for $\displaystyle\int_0^b \cos x \, dx$	b	Riemann sum for $\displaystyle\int_0^b \cos x \, dx$
0.0	0	3.5	-0.35
0.5	0.48	4.0	
1.0		4.5	
1.5		5.0	
2.0		5.5	
2.5		6.0	
3.0		6.5	

TABLE 1

[†]Riemann sum program 3B is used on this worksheet to predict a formula for an integral of $\cos x$ and to predict relationships among integrals of $\sin(\pi x)$ over different intervals. See Calculator instructions 1B.3 for tips on working it with a calculator.

$y = \int_0^b \cos x \, dx$

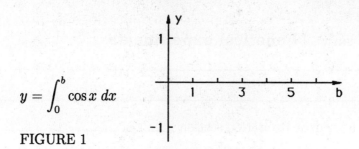

FIGURE 1

Problem 2a Use program 3B to find the Riemann sum for $\int_0^{0.5} \sin(\pi x) \, dx$ with two subintervals ($N = 2$) and the function evaluated at the midpoints ($T = 0.5$). Use $-0.5 \leq x \leq 3$ and $-1.5 \leq y \leq 1.5$ for the graph, and copy the curve and rectangles in Figure 2.

b. Find the Riemann sum for $\int_0^1 \sin(\pi x) \, dx$ with $N = 4$ and $T = 0.5$. Copy the curve and rectangles in Figure 3.

c. Find the Riemann sum for $\int_0^{2.5} \sin(\pi x) \, dx$ with $N = 10$ and $T = 0.5$. Copy the curve and rectangles in Figure 4.

d. Use the drawings to explain why the Riemann sum in part (b) is twice the Riemann sum in part (a) and why the Riemann sum in part (c) equals the Riemann sum in part (a).

e. Find the Riemann sums for $\int_0^{0.5} \sin(\pi x) \, dx$ with $N = 25$ and $T = 0.5$ and for $\int_0^1 \sin(\pi x) \, dx$ with $N = 50$ and $T = 0.5$. Why do you think the second Riemann sum is equal to twice the first? What does this and the result of part (d) suggest about the two integrals?

f Find the Riemann sum for $\int_0^{2.5} \sin(\pi x) \, dx$ with $N = 125$ and $T = 0.5$ Why do you think this Riemann sum is equal to the first Riemann sum from part (e)? What does this and the result of part (d) suggest about the two integrals?

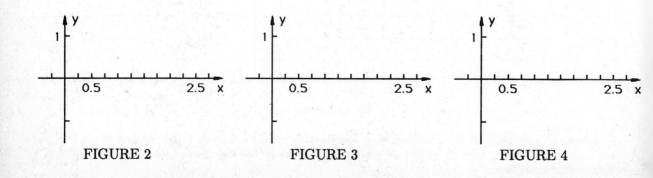

FIGURE 2　　　　　　　FIGURE 3　　　　　　　FIGURE 4

Graphing calculator workbook

Worksheet 3C.3[†]

Algebra of integrals: Numerical experiments

Name _____ Date _____

Others in your group _____

Instructor, Teaching Assistant, and/or Recitation section _____

Directions Put first drafts of your calculations and answers on scratch paper. Take your time, work carefully, and discuss your solution with at least one other student before putting a final draft on this sheet or on other paper. Turn in all your work.

Recall that a Midpoint rule approximation is a Riemann sum with the integrand evaluated at the midpoints of equal subintervals, and are obtained by using $T = 0.5$ in Program 3B. Round off results to four decimal places in these problems.

Problem 1a Calculate the Midpoint rule approximation of $\int_0^1 (4 - 7x^2 + 3x^3)\,dx$ corresponding to 5 subintervals. (Use $-0.2 \leq x \leq 2.2$, $-5 \leq y \leq 9$ and set $A = 0$, $B = 1$, $N = 5$, and $T = 0.5$ in program 3B.) Copy the curve and rectangles and put the value of the Riemann sum in Figure 1.

b. Find the Midpoint rule approximation of $\int_1^2 (4 - 7x^2 + 3x^3)\,dx$ corresponding to 5 subintervals. (Use the ranges from part (a), $A = 1$, $B = 2$, $N = 5$, and $T = 0.5$.) Copy the curve and rectangles and put the value of the Riemann sum in Figure 2.

c. Find the Midpoint rule approximation of $\int_0^2 (4 - 7x^2 + 3x^3)\,dx$ corresponding to 10 subintervals. (Use the ranges from part (a), $A = 0$, $B = 2$, $N = 10$, and $T = 0.5$.) Copy the curve and rectangles and put the value of the Riemann sum in Figure 3.

d. How are the Riemann sums in parts (a), (b), and (c) related? Explain.

e. Use program 3B to find the Midpoint rule approximation $\int_0^1 (4 - 7x^2 + 3x^3)\,dx$ and $\int_1^2 (4 - 7x^2 + 3x^3)\,dx$ with 20 subintervals and of $\int_0^2 (4 - 7x^2 + 3x^3)\,dx$ with 40 subintervals. How are they related? Explain.

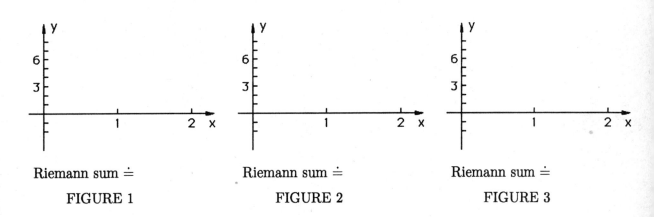

Riemann sum ≐   Riemann sum ≐   Riemann sum ≐

FIGURE 1   FIGURE 2   FIGURE 3

[†]This worksheet uses Riemann sum program 3B to predict integration rules. See Calculator instructions 1B.3 for tips on working it with a calculator.

f. How would the Midpoint rule approximations of $\int_0^1 (4 - 7x^2 + 3x^3)\, dx$ and $\int_1^2 (4 - 7x^2 + 3x^3)\, dx$ with N subintervals be related to the Midpoint rule approximations of $\int_0^2 (4 - 7x^2 + 3x^3)\, dx$ with $2N$ subintervals for any positive integer N? What does this say about the integrals themselves?

Problem 2 Find the Midpoint rule approximation of $\int_0^2 2(4 - 7x^2 + 3x^3)\, dx$ with 10 subintervals. Use the ranges from Problem 1. Put the curve, the rectangles, and the value of the Riemann sum in Figure 4. How is this sum related to the Riemann sum of Figure 3? How are the corresponding integrals related?

Problem 3 Find the Midpoint rule approximation of $\int_0^2 -(4 - 7x^2 + 3x^3)\, dx$ with 10 equal subintervals and the ranges from Problem 1. Put the curve, the rectangles, and the value of the Riemann sum in Figure 5. How is this sum related to the Riemann sum of Figure 3? How are the corresponding integrals related?

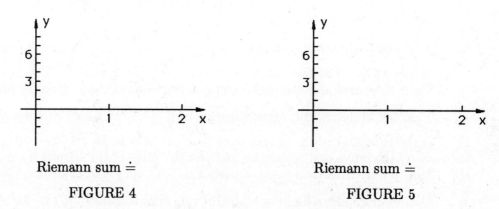

Riemann sum $\doteq$

FIGURE 4

Riemann sum $\doteq$

FIGURE 5

Problem 4 Set x-min $= -0.2$, x-max $= 2.2$, y-min $= -1$, and y-max $= 8$. Find the Midpoint rule approximations of $\int_0^2 (x+1)\, dx$, $\int_0^2 3\sqrt{x}\, dx$, and $\int_0^2 (x + 1 + 3\sqrt{x})\, dx$ with six subintervals. Put the curve, the rectangles, and the value of the Riemann sum in Figures 6 through 8. How are these Riemann sums related? How are the corresponding integrals related?

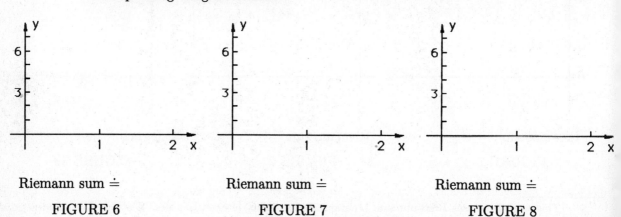

Riemann sum $\doteq$

FIGURE 6

Riemann sum $\doteq$

FIGURE 7

Riemann sum $\doteq$

FIGURE 8

Worksheet 3D.1[†]

The Fundamental theorem: A geometric interpretation

Name _____ Date _____

Others in your group _____

Instructor, Teaching Assistant, and/or Recitation section _____

Directions *Put your initial calculations and first drafts of your conclusions on scratch paper. Take your time, work carefully, and give your conclusions in well written sentences. Discuss your solution with at least one other student before putting a final draft on this sheet or other sheets of paper, as needed. Turn in all your work.*

Problem 1 Suppose that after you have painted a circle of radius r, you decide to make it slightly larger so its radius is $r + \Delta r$ (Figure 1). Let A denote the area πr^2 of the original circle and let ΔA denote the increase in area for the larger circle. Give a geometric explanation of why

$$\Delta A \approx 2\pi r \Delta r \quad \text{for small } \Delta r.$$

Problem 2 Suppose that you want to enlarge a square of width w into a square of width $w + \Delta w$ as in Figure 2. Let A denote the area w^2 of the original square and let ΔA denote the increase in area for the larger square. Give a geometric explanation of why

$$\Delta A \approx 2w\Delta w \quad \text{for small } \Delta w.$$

Problem 3 Let A denote the area of the rectangle of width x and height 2 in Figure 3. Show that the increase ΔA of the area of the rectangle that is obtained when the width is increased from x to to $x + \Delta x$ is given by

$$\Delta A = 2\Delta x. \tag{1}$$

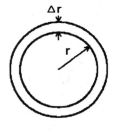

FIGURE 1

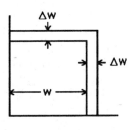

FIGURE 2

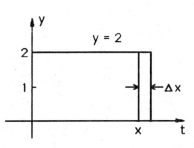

FIGURE 3

[†]This worksheet looks at examples where making a small increase Δx in one dimension of a geometric figure increases the area by an amount that is approximately equal to Δx multiplied by the length of the portion of the boundary of the region that is moved. This gives geometric meaning to the Fundamental theorem for differentiating integrals with respect to their upper limits. The problems do not use any integration formulas or prior knowledge of the Fundamental theorem.

Problem 4 Let A denote the area of the triangle of width x in Figure 4 and let ΔA be the increase in area that is obtained when the width is increased to $x + \Delta x$ Give a geometric explanation of why

$$\Delta A \approx x \Delta x \quad \text{for small } \Delta x. \tag{2}$$

Problem 5 Figure 5 shows the region above the t-axis and below the graph of the function t^2 for $0 \leq t \leq x$ with a positive x, and the region that is obtained by increasing the width to $x + \Delta x$. Give a geometric explanation of why the increase ΔA in the area satisfies

$$\Delta A \approx x^2 \Delta x \quad \text{for small } \Delta x. \tag{3}$$

Problem 6 Statements (1), (2), and (3) with the fact that

$$\frac{dA}{dx} = \lim_{\Delta x \to 0} \frac{\Delta A}{\Delta x}$$

illustrate the Fundamental theorem of calculus for derivatives of integrals, which is stated below. Explain.

The Fundamental theorem for derivatives of integrals

Hypotheses: $f(t)$ is continuous in an interval I containing the point a.

Conclusion: For x in the interior of I, $\dfrac{d}{dx} \displaystyle\int_a^x f(t)\, dt = f(x)$.

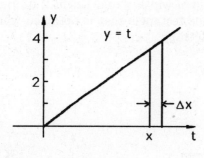

FIGURE 4

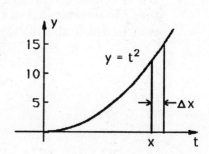

FIGURE 5

Graphing calculator workbook

Worksheet 3D.2[†]

The Fundamental theorem: Examples and nonexamples

Name _____ Date _____

Others in your group _____

Instructor, Teaching Assistant, and/or Recitation section _____

Directions Put first drafts of your calculations and answers on scratch paper. Take your time, work carefully, and discuss your solution with at least one other student before putting a final draft on this sheet or on other paper. Turn in all your work.

The Fundamental theorem for integrals of derivatives

Hypotheses: $F(x)$ has a continuous derivative in an interval containing a and b.

Conclusion: $\int_a^b F'(x)\,dx = F(b) - F(a)$.

Problem 1 Sketch the graph of $F_1(x) = 2x^2 + 10$ in Figure 1 with $-1 \leq x \leq 5, -10 \leq y \leq 50$ and the graph of its derivative $F_1'(x)$ in Figure 2 with $-1 \leq x \leq 5, -5 \leq y \leq 25$.

(a) Calculate $F_1(4) - F_1(0)$. (b) Find $\int_0^4 F_1'(x)\,dx$. (c) Does $F_1(x)$ satisfy the hypotheses of the Fundamental theorem stated above with $a = 0$ and $b = 4$? If not, why not? Does it satisfy the conclusion? If not, why not?

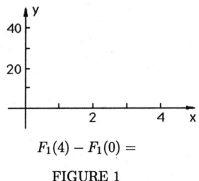

$F_1(4) - F_1(0) =$

FIGURE 1

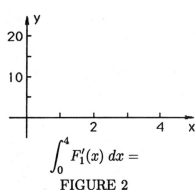

$\int_0^4 F_1'(x)\,dx =$

FIGURE 2

Problem 2 Follow the directions of Problem 1 with $F_2(x) = \begin{cases} 16x - 2x^2 & \text{for } x \leq 2 \\ 8 + 8x & \text{for } x > 2. \end{cases}$

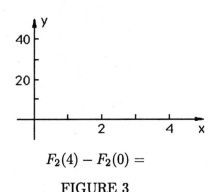

$F_2(4) - F_2(0) =$

FIGURE 3

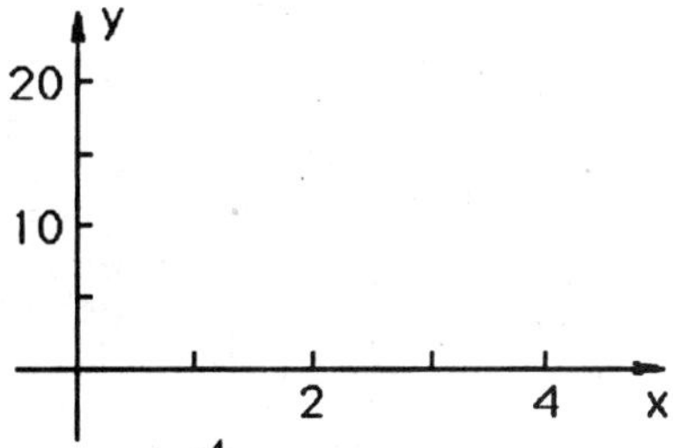

$\int_0^4 F_2'(x)\,dx =$

FIGURE 4

[†]This worksheet studies examples in which the Fundamental theorem for integrals of derivatives can and cannot be applied. See Calculator instructions 1B.3 for tips on working it with a calculator.

Problem 3 Follow the directions of Problem 1 with $F_3(x) = \begin{cases} 15x & \text{for} \quad x \leq 2 \\ 5x + 20 & \text{for} \quad x > 2. \end{cases}$

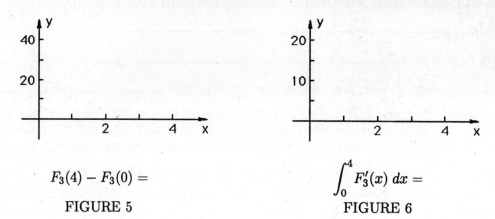

$F_3(4) - F_3(0) =$

FIGURE 5

$\int_0^4 F_3'(x)\, dx =$

FIGURE 6

Problem 4 Follow the directions of Problem 1 with $F_4(x) = \begin{cases} 20x & \text{for} \quad x \leq 2 \\ 3x^2 & \text{for} \quad x > 2. \end{cases}$

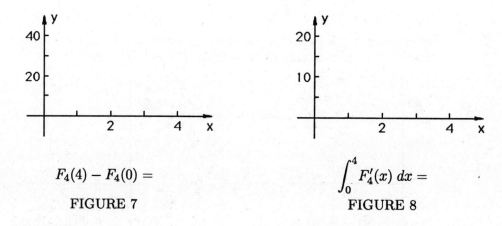

$F_4(4) - F_4(0) =$

FIGURE 7

$\int_0^4 F_4'(x)\, dx =$

FIGURE 8

Problem 5 Follow the directions of Problem 1 with $F_5(x) = \begin{cases} 4 & \text{for} \quad x = 0 \\ 3x^2 + 20 & \text{for} \quad 0 < x \leq 2 \\ 18x - 20 & \text{for} \quad x > 2. \end{cases}$

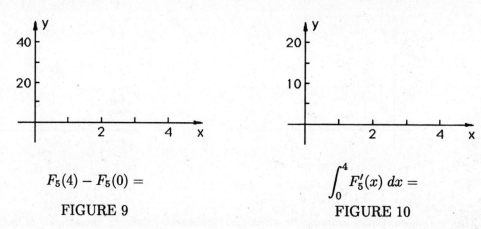

$F_5(4) - F_5(0) =$

FIGURE 9

$\int_0^4 F_5'(x)\, dx =$

FIGURE 10

Graphing calculator workbook

Simpson's rule program 3E

Texas Instruments TI-81

A Simpson's rule approximation of the integral $\int_a^b f(x)\,dx$ with $a < b$ is obtained by dividing the interval $[a, b]$ into a even number of subintervals and approximating the graph of $f(x)$ over each pair of subintervals by a parabola. We let M denote the number of pairs of subintervals, so that the number of subintervals is $2M$ and the corresponding partition of $[a, b]$ is

$$a = x_0 < x_1 < x_2 < \cdots < x_{2M-1} < x_{2M} = b.$$

The parabola that approximates the graph of $f(x)$ for x in the kth pair of subintervals intersects the graph at $x = x_{2k-2}, x = x_{2k-1}$, and $x = x_{2k}$. A straight forward but lengthy calculation shows that the resulting approximation of the portion of the integral from x_{2k-2} to x_{2k} is

(1) $$\tfrac{1}{3}\left[f(x_{2k-2}) + 4f(x_{2k-1}) + f(x_{2k})\right]\Delta x$$

with $\Delta x = (b - a)/(2M)$ the width of the subintervals. The approximation of the integral is the sum of the numbers (1):

(2) $$\text{Simpson's rule approximation} = \sum_{k=1}^{M} \tfrac{1}{3}\left[f(x_{2k-2}) + 4f(x_{2k-1}) + f(x_{2k})\right]\Delta x.$$

The program given below calculates this approximation for the function Y_1 in the $Y =$ menu of the calculator. The user picks suitable ranges of x and y for the graph, the limits of integration a and b, and an even number N of subintervals to be used. The program draws the graph, vertical lines from the x-axis to the graph at the endpoints of the subintervals, and gives the approximate decimal value of the approximation.

The program (250 bytes)

1. Prgm4: SIMPSON
2. : ClrDraw
3. : All-Off
4. : Y_1-On
5. : Disp "USE $Y_1(X)$"
6. : Disp "$A =$"
7. : Input A
8. : Disp "$B =$"
9. : Input B
10. : Disp "NUMBER N OF"
11. : Disp "INTERVALS"
12. : Input N
13. : $N/2 \to M$

14.	: If Int $M = M$
15.	: Goto 1
16.	: Disp "USE EVEN N"
17.	: Stop
18.	: Lbl 1
19.	: $(B - A)/N \to D$ (Note: Use $\boxed{-}$ not $\boxed{(-)}$.)
20.	: $\emptyset \to S$ (Note: $\emptyset$ denotes zero.)
21.	: $1 \to K$
22.	: Lbl 2
23.	: $A + 2(K - 1)D \to P$
24.	: $P \to X$
26.	: $Y_1 \to U$
26.	: $P + D \to X$
27.	: $Y_1 \to V$
28.	: $P + 2D \to X$
29.	: $Y_1 \to W$
30.	: Line$(P, \emptyset, P, U)$
31.	: Line$(P + D, \emptyset, P + D, V)$
32.	: Line$(P + 2D, \emptyset, P + 2D, W)$
33.	: $S + D(U + 4V + W)/3 \to S$
34.	: $K + 1 \to K$
35.	: If $K \le M$
36.	: Goto 2
37.	: Pause
38.	: Disp "SIMPSON APPROX ="
39.	: Disp S
40.	: $S + \emptyset$

Entering the program

See the instructions for program 2B or 3B. Recall that $\emptyset$ denotes zero.

---

1. Press $\boxed{S}$ $\boxed{I}$ $\boxed{M}$ $\boxed{P}$ $\boxed{S}$ $\boxed{O}$ $\boxed{N}$ $\boxed{\text{ENTER}}$. The calculator is automatically in alpha mode here. This title identifies the program.

---

2. Press $\boxed{\text{2nd}}$ $\boxed{\text{DRAW}}$ $\boxed{1}$ $\boxed{\text{ENTER}}$. ClrDraw clears the graphics screen.

---

3. Press $\boxed{\text{2nd}}$ $\boxed{\text{Y-VARS}}$ $\boxed{\blacktriangleleft}$ $\boxed{1}$ $\boxed{\text{ENTER}}$. All-Off unselects the four functions in the $Y =$ menu so their graphs will not be generated.

Graphing calculator workbook

4. Press [2nd] [Y-VARS] [▶] [2] [ENTER]. This selects Y_1.

5. Press [PRGM] [▶] [1] [2nd] [A-LOCK] ["] [U] [S] [E] [_] [2nd] [Y-VARS] [1] [(] [X|T] [)] [ALPHA] ["] [ENTER] with [_] the space (on the 0 key).

6. Press [PRGM] [▶] [1] [ALPHA] ["] [ALPHA] [A] [2nd] [TEST] [1] [ALPHA] ["] [ENTER].

7. Press [PRGM] [▶] [2] [ALPHA] [A] [ENTER].

8. Press [PRGM] [▶] [1] [ALPHA] ["] [ALPHA] [B] [2nd] [TEST] [1] [ALPHA] ["] [ENTER].

9. Press [PRGM] [▶] [2] [ALPHA] [B] [ENTER].

10. Press [PRGM] [▶] [1] [2nd] [A-LOCK] ["] [N] [U] [M] [B] [E] [R] [_] [N] [_] [O] [F] ["] [ENTER].

11. Press [PRGM] [▶] [1] [2nd] [A-LOCK] ["] [I] [N] [T] [E] [R] [V] [A] [L] [S] ["] [ENTER].

12 Press [PRGM] [▶] [2] [ALPHA] [N] [ENTER].

13. Press [ALPHA] [N] [÷] [2] [STO▶] [M] [ENTER]. The number of pairs of subintervals $M = N/2$ is calculated and stored.

14. Press [PRGM] [3] [MATH] [▶] [4] [ALPHA] [M] [2nd] [TEST] [1] [ALPHA] [M] [ENTER]. If N is even, the next command is executed.

15. Press [PRGM] [2] [1] [ENTER].

16. Press [PRGM] [▶] [1] [2nd] [A-LOCK] [”] [U] [S] [E] [_] [E] [V] [E] [N] [_] [N] [”] [ENTER].

17. Press [PRGM] [8] [ENTER]. If N is odd, the previous line explains what has happened and this line terminates the program.

18. Press $\boxed{\text{PRGM}}$ $\boxed{1}$ $\boxed{1}$ $\boxed{\text{ENTER}}$.

19. Press $\boxed{(}$ $\boxed{\text{ALPHA}}$ $\boxed{B}$ $\boxed{-}$ $\boxed{\text{ALPHA}}$ $\boxed{A}$ $\boxed{)}$ $\boxed{\div}$ $\boxed{\text{ALPHA}}$ $\boxed{N}$ $\boxed{\text{STO}\blacktriangleright}$ $\boxed{D}$ $\boxed{\text{ENTER}}$. The width D of the subintervals is calculated and stored.

20. Press $\boxed{\emptyset}$ $\boxed{\text{STO}\blacktriangleright}$ $\boxed{S}$ $\boxed{\text{ENTER}}$. The sum is stored as S during the calculations.

21. Press $\boxed{1}$ $\boxed{\text{STO}\blacktriangleright}$ $\boxed{K}$ $\boxed{\text{ENTER}}$. $K = 1$ for the first pair of intervals.

22. Press $\boxed{\text{PRGM}}$ $\boxed{1}$ $\boxed{2}$ $\boxed{\text{ENTER}}$. This loop ends on line 36.

23. Press $\boxed{\text{ALPHA}}$ $\boxed{A}$ $\boxed{+}$ $\boxed{2}$ $\boxed{(}$ $\boxed{\text{ALPHA}}$ $\boxed{K}$ $\boxed{-}$ $\boxed{1}$ $\boxed{)}$ $\boxed{\text{ALPHA}}$ $\boxed{D}$ $\boxed{\text{STO}\blacktriangleright}$ $\boxed{P}$ $\boxed{\text{ENTER}}$. x_{2K-2} is stored as P.

24. Press $\boxed{\text{ALPHA}}$ $\boxed{P}$ $\boxed{\text{STO}\blacktriangleright}$ $\boxed{X|T}$ $\boxed{\text{ENTER}}$.

25. Press $\boxed{\text{2nd}}$ $\boxed{\text{Y-VARS}}$ $\boxed{1}$ $\boxed{\text{STO}\blacktriangleright}$ $\boxed{U}$ $\boxed{\text{ENTER}}$. $U = Y_1(x_{2K-2})$.

26. Press $\boxed{\text{ALPHA}}$ $\boxed{P}$ $\boxed{+}$ $\boxed{\text{ALPHA}}$ $\boxed{D}$ $\boxed{\text{STO}\blacktriangleright}$ $\boxed{X|T}$ $\boxed{\text{ENTER}}$. The common endpoint x_{2K-1} of the two intervals is stored as X.

27. Press $\boxed{\text{2nd}}$ $\boxed{\text{Y-vars}}$ $\boxed{1}$ $\boxed{\text{STO}\blacktriangleright}$ $\boxed{V}$ $\boxed{\text{ENTER}}$. $V = Y_1(x_{2K-1})$.

28. Press $\boxed{\text{ALPHA}}$ $\boxed{P}$ $\boxed{+}$ $\boxed{2}$ $\boxed{\text{ALPHA}}$ $\boxed{D}$ $\boxed{\text{STO}\blacktriangleright}$ $\boxed{X|T}$ $\boxed{\text{ENTER}}$. The right endpoint x_{2K} of the two intervals is stored as X.

29. Press $\boxed{\text{2nd}}$ $\boxed{\text{Y-vars}}$ $\boxed{1}$ $\boxed{\text{STO}\blacktriangleright}$ $\boxed{W}$ $\boxed{\text{ENTER}}$. $W = Y_1(x_{2K})$.

30. Press $\boxed{\text{2nd}}$ $\boxed{\text{DRAW}}$ $\boxed{2}$ $\boxed{\text{ALPHA}}$ $\boxed{P}$ $\boxed{\text{ALPHA}}$ $\boxed{,}$ $\boxed{\emptyset}$ $\boxed{\text{ALPHA}}$ $\boxed{,}$ $\boxed{\text{ALPHA}}$ $\boxed{P}$ $\boxed{\text{ALPHA}}$ $\boxed{,}$ $\boxed{\text{ALPHA}}$ $\boxed{U}$ $\boxed{)}$ $\boxed{\text{ENTER}}$. This and the next two commands draw vertical lines from the x-axis to the graph.

31. Press $\boxed{\text{2nd}}$ $\boxed{\text{DRAW}}$ $\boxed{2}$ $\boxed{\text{ALPHA}}$ $\boxed{P}$ $\boxed{+}$ $\boxed{\text{ALPHA}}$ $\boxed{D}$ $\boxed{\text{ALPHA}}$ $\boxed{,}$ $\boxed{\emptyset}$ $\boxed{\text{ALPHA}}$ $\boxed{,}$ $\boxed{\text{ALPHA}}$ $\boxed{P}$ $\boxed{+}$ $\boxed{\text{ALPHA}}$ $\boxed{D}$ $\boxed{\text{ALPHA}}$ $\boxed{,}$ $\boxed{\text{ALPHA}}$ $\boxed{V}$ $\boxed{)}$ $\boxed{\text{ENTER}}$.

Graphing calculator workbook

32. Press [2nd] [DRAW] [2] [ALPHA] [P] [+] [2] [ALPHA] [D]
[ALPHA] [,] [0] [ALPHA] [,] [ALPHA] [P] [+] [2] [ALPHA]
[D] [ALPHA] [,] [ALPHA] [W] [)] [ENTER].

33. Press [ALPHA] [S] [+] [ALPHA] [D] [(] [ALPHA] [U] [+]
[4] [ALPHA] [V] [+] [ALPHA] [W] [)] [÷] [3] [STO▶] [S]
[ENTER]. The contribution from the Kth pair of subintervals is added to S.

34. Press [ALPHA] [K] [+] [1] [STO▶] [K] [ENTER]. Increase K by 1.

35. Press [PRGM] [3] [ALPHA] [K] [2nd] [TEST] [6] [ALPHA]
[M] [ENTER]. If K is $\leq M$, the loop is repeated for the next pair of intervals.

36 Press [PRGM] [2] [2] [ENTER].

37. Press [PRGM] [6] [ENTER].

38. Press [PRGM] [▶] [1] [2nd] [A-LOCK] [“] [S] [I] [M] [P]
[S] [O] [N] [–] [A] [P] [P] [R] [O] [X] [2nd] [TEST]
[1] [ALPHA] [“] [ENTER]. The symbols SIMPSON APPROX = are displayed.

39. Press [PRGM] [▶] [1] [ALPHA] [S] [ENTER]. The value of the Simpson's
rule approximation is displayed.

39. Press [ALPHA] [+] [0]. The approximation is stored as *Ans* for subsequent calculations.

Using the program

Example 1 The integral $\int_0^3 [1 + x + x\sin(x^2)]\ dx$ has the exact value $8 - \frac{1}{2}\cos(9)$. Compare this with the values obtained from Simpson's rule and the midpoint rule with $10, 20$ and 40 subintervals.

Solution Put $Y_1 = 1 + X + X\sin(X^2)$ and set x-min $= -0.5$, x-max $= 3.5$ to include the interval $[0, 3]$ of integration. After some experimentation, we pick $-1 \leq y \leq 7$ for the y-range. Running the Simpson's rule program with $A = 0$, $B = 3$, and $N = 10$ gives the graph in Figure 1 and the value 8.477890343, which is stored as S when the program has finished. (The sum is not the value ANS that is recalled with [ENTER] or [▲] because it comes from a program command rather than being the last result of operations in the home screen.)

Evaluate the integral:

$$\int_0^3 ([1 + x + x\sin(x^2)])\, dx = \left[x + \tfrac{1}{2}x^2 - \tfrac{1}{2}\cos(x^2)\right]_0^3 = 8 - \tfrac{1}{2}\cos(9).$$

Press $\boxed{8}$ $\boxed{-}$ $\boxed{(}$ $\boxed{\cos}$ $\boxed{9}$ $\boxed{)}$ $\boxed{\div}$ $\boxed{2}$ $\boxed{\text{STO}\triangleright}$ $\boxed{I}$ $\boxed{\text{ENTER}}$ to store the value of the integral as I. Then press $\boxed{\text{ALPHA}}$ $\boxed{S}$ $\boxed{-}$ $\boxed{\text{ALPHA}}$ $\boxed{I}$ $\boxed{\text{ENTER}}$ to see that the error is approximately 2.23×10^{-2}. Run the Simpson's rule program with $N = 20$ and 40 and the Riemann sum program 3B with $N = 10, 20$, and 40 and $T = 0.5$ for the other values in the table.

Notice that Simpson's rule is more accurate than the midpoint rule with the same number of subintervals. In fact, quadrupling the number of subintervals from 10 to 40 decreases the error in Simpson's rule by the factor $(2.23 \times 10^{-2})/(8.73 \times 10^{-5}) \doteq 255$ but decreases the error with the midpoint rule only by the factor $(6.50 \times 10^{-2})/(3.77 \times 10^{-3}) \doteq 17$. This reflects the fact that the error with Simpson's rule is bounded by a constant times $(\Delta x)^4$, while the error with the midpoint rule is bounded by a constant times $(\Delta x)^2$.

	Simpson's rule		The midpoint rule	
N	Sum	Error	Sum	Error
10	8.477890343	2.23×10^{-2}	8.520564394	6.50×10^{-2}
20	8.457014063	1.45×10^{-3}	8.470859221	1.53×10^{-2}
40	8.455652446	8.73×10^{-5}	8.459331269	3.77×10^{-3}

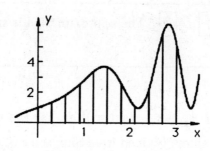

FIGURE 1

Graphing calculator workbook   157

Construction 3F.1[†]

Volumes by slicing: Three-dimensional models

Name _____ Date _____

Others in your group _____

Instructor, Teaching Assistant, and/or Recitation section _____

Directions   Put first drafts of your calculations and answers on scratch paper. Take your time, work carefully, and discuss your solution with at least one other student before putting a final draft on this sheet or on other paper. Turn in all your work.

Problem 1   Slice this page along the horizontal dashed line in the figure below. Cut out the rectangle, five semicircles, and five equilateral triangles on the next page and reinforce them with transparent tape. Make hill folds along the three horizontal thin dashed lines labeled A, B, and C in the rectangle and valley folds along the two heavy dashed lines labeled D and E. Slice along the five solid, vertical lines inside the rectangle. Fold the rectangle along line B and slide the middle section from the back through the slice on this page. Fold and tape the upper and lower sections to the back of this sheet. Re-establish the hill folds to form a support for the semicircles and triangles.

a.   The curves in Figure 1 are $y = 1 - x + \frac{1}{3}x^2$ and $y = 1 + x - \frac{1}{3}x^2$. Place the five semicircles in the support according to the x-values printed on them to form a model of the solid whose base is the region between these curves and whose cross sections perpendicular to the x-axis are semicircles with diameters on the xy-plane. Use the method of slicing to show that the volume of this solid is $\frac{9}{20}\pi$.

b.   Replace the semicircles by the equilateral triangles. Then calculate the volume of the solid with triangular cross sections that this model represents. Express it as a fraction multiplied by $\sqrt{3}$.

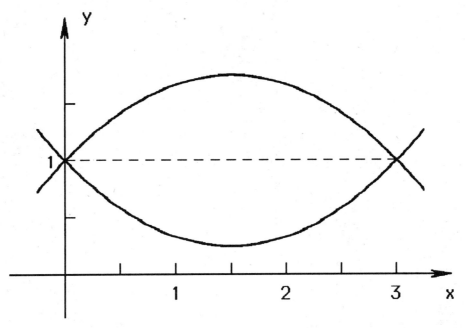

[†]This construction illustrates what it means for cross sections of a solid perpendicular to an x-axis to be semicircles or triangles. The volumes can be found by the method of slicing.

Graphing calculator workbook

Graphing calculator workbook 159

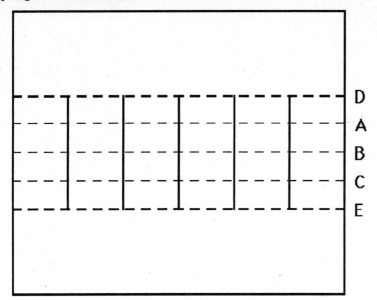
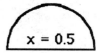
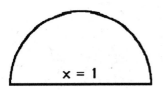
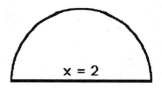
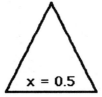
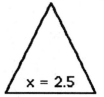
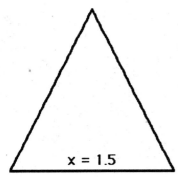
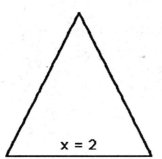
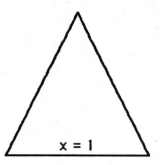

Graphing calculator workbook

Graphing calculator workbook 161

Construction 3F.2[†]

Solids of revolution: Three-dimensional models

Name _____ Date _____

Others in your group _____

Instructor, Teaching Assistant, and/or Recitation section _____

Directions Put first drafts of your calculations and answers on scratch paper. Take your time, work carefully, and discuss your solution with at least one other student before putting a final draft on this sheet or on other paper. Turn in all your work.

Problem 1 Figure 1 shows the region R between the x-axis and the curve $y = \sqrt{x-1}$ for $1 \leq x \leq 3$. Slice this sheet along the right side of the region and the four dashed vertical lines. Cut out the four semicircles with rectangular tabs on the next page by cutting along the solid lines. Reinforce them with transparent tape, and fold the tabs in opposite directions along the dashed lines. Put the semicircles through this sheet from the back at the indicated values of x and tape the tabs to the back. Stand the semicircles up to form a model of half of the solid that is obtained by rotating the region around the x-axis. Use the method of slicing to find the exact volume of the entire solid, whose cross sections perpendicular to the x-axis are circles. Express it as a fraction multiplied by π.

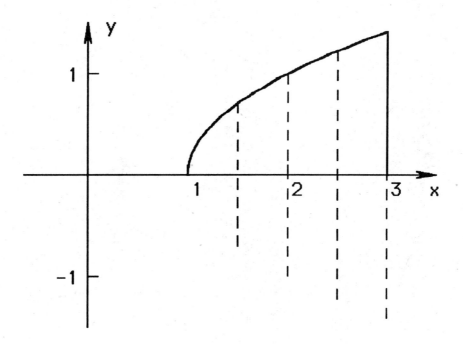

FIGURE 1

[†]These constructions illustrate solids generated by rotating regions about the x- and y-axes. The volumes in both problems can be found by the method of slicing. The volume in Problem 2 can also be determined by the method of cylindrical shells.

Graphing calculator workbook

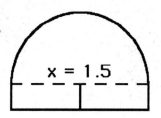

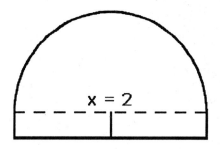

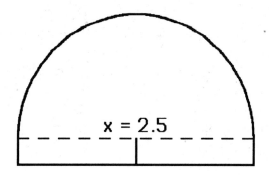

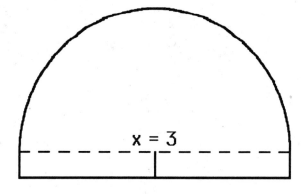

Graphing calculator workbook

Graphing calculator workbook 165

Problem 2 Figure 2 shows eight copies of the region from Problem 1 placed in a circle. Use a knife to slice the tops and sides of each region along the solid lines. Fold up the regions along the dashed lines to make a model of the solid that is obtained by rotating the region around the y-axis. The y-axis is represented by the dot; it would be perpendicular to this page. Use the method of slicing with a y-integral and the equation $x = y^2 + 1, 0 \leq y \leq \sqrt{2}$ for the curved top of the region to show that the volume of the solid is $\frac{88}{15}\pi\sqrt{2}$. (You could also use $y = \sqrt{x-1}$ and the method of cylindrical shells if you know that technique.)

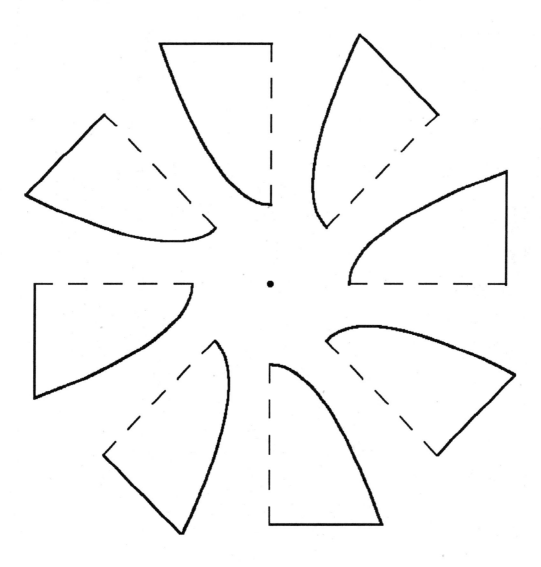

FIGURE 2

Graphing calculator workbook

Project 3G.1[1]

Designing a shish-kabob

Name _____ Date _____

Others in your group _____

Instructor, Teaching Assistant, and/or Recitation section _____

Directions Put first drafts of your calculations and answers on scratch paper. Take your time, work carefully, and discuss your solution with at least one other student before putting a final draft on this sheet or on other paper. Turn in all your work.

Problem 1 You could describe a shish-kabob (Figure 1) by defining the cross sections of the skewer and of each piece of meat and vegetable in planes perpendicular to the skewer. Then formulas for the areas of the cross sections of the skewer and of the pieces of food as functions of x, measured along the skewer, could be used to find the x-coordinate of the center of gravity of the shish-kabob.

Define such a shish-kabob using at least two differently shaped pieces of meat, one tomato, and one mushroom and with a complete or partial circle at the end of the skewer. Use spheres to represent tomatoes, cylinders for meat, and cylinders topped by hemispheres for mushrooms. Then adjust the positions of the food to have the shish-kabob balance at the center of the skewer. Use 4 grams per cubic centimeter for the density of the skewer, 2 grams per cubic centimeter for meat, and 1.5 grams per cubic centimeter for tomatoes and mushrooms.

FIGURE 1

---

[1]This project involves finding centers of gravity by the method of slicing.

Graphing calculator workbook

Graphing calculator workbook

Project 3G.2[†]

Billy the Kid

Name _____ Date _____

Others in your group _____

Instructor, Teaching Assistant, and/or Recitation section _____

Directions   Put first drafts of your calculations and answers on scratch paper. Take your time, work carefully, and discuss your solution with at least one other student before putting a final draft on this sheet or on other paper. Turn in all your work.

Problem 1   In 1878, the nineteeen year old Billy Bonney (the Kid) and a former buffalo hunter Pat Garrett were in the territory of New Mexico. Imagine that at the same time one morning they set out on horseback. Billy starts near the intersection of the Pecos River and the Taiban Creek at the point labeled B in the map on the back of this sheet. He rides in a straight line to the town of Stinking Springs. Pat starts at point G and rides north to the Wilcox-Brazil ranch. Near point J they are close enough to recognize each other. (They were acquainted; Pat's sister-in-law, Celsa Gutierrez, is said to have been one of Billy's lovers.) They do not realize that three years later Pat, as sheriff of Lincoln County, would arrest Billy near Stinking Springs on a murder charge, that Billy would escape from custody in Lincoln, and that Pat would eventually track him down and shoot him in a dark bedroom in Fort Sumner.

   a.   Define functions $s_B(t)$ and $s_P(t)$ of the time t so that $s_B(t)$ is Billy's speed as he rides from point B to Stinking Springs and $s_P(t)$ is Pat's speed as he rides from point G to the Wilcox-Brazil ranch. Choose these functions so that they each ride part of the time at a walk (4 miles per hour), part of the time at a trot (9 miles per hour), and part of the time at a canter (11 miles per hour), and so they pass within 50 feet of each other.

Billy the Kid[1]
(1859–1881)

Pat Garrett[2]
(1850–1908)

[†]This worksheet requires finding distances from velocities.
[1] From a photograph, by special permission of the Lincoln County Heritage Trust
[2] From a photograph, courtesy Leon C. Metz

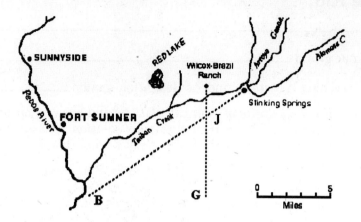

FIGURE 1

b. Put rectangular coordinates on the map and use them with approximate measurements of distances to find formulas for Billy's and Pat's coordinates as functions of t on their rides and then to express the distance between Billy and Pat as a function of t.

Graphing calculator workbook

Project 3G.3[†]

Designing a boat

Name _____ Date _____

Others in your group _____

Instructor, Teaching Assistant, and/or Recitation section _____

Directions *Put first drafts of your calculations and answers on scratch paper. Take your time, work carefully, and discuss your solution with at least one other student before putting a final draft on this sheet or on other paper. Turn in all your work.*

Problem 1 Your objective in this worksheet is to design a boat. Not a real boat (Figure 1), but a simplified fiberglass rowboat (Figure 2) with a horizontal top, flat horizontal bottom, and vertical sides.

Decide how large you want the boat to be and the weight of passengers and cargo you want it to carry. Pick functions whose graphs give a shape you like for its top and bottom (Figure 3) and make a preliminary choice of the height of its sides.

Determine how high the top of the boat would be above the water when it is fully loaded (or whether it would sink) by using Archimedes' principle that a boat displaces its weight in water. Assume that water weighs 62.4 pounds per cubic foot and that the combination of fiberglass and resin that is used to build the sides and bottom of the boat weighs 0.75 pounds per square foot.

Redesign the boat if necessary so that the water line is between one foot and eighteen inches below the top of the boat when it is fully loaded.

FIGURE 1

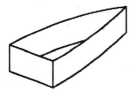

FIGURE 2

FIGURE 3

[†]This worksheet uses Archimedes' principle and the method of slicing, applied to finding weights.

Graphing calculator workbook

Graphing calculator workbook

Worksheet 4A.1†

Derivatives of exponential functions: Numerical experiments

Name _____ Date _____

Others in your group _____

Instructor, Teaching Assistant, and/or Recitation section _____

Directions Put first drafts of your calculations and answers on scratch paper. Take your time, work carefully, and discuss your solution with at least one other student before putting a final draft on this sheet or on other paper. Turn in all your work.

Problem 1a Generate the graphs of 2^x and 3^x for $-1.5 \leq x \leq 2, -1 \leq y \leq 5$ on your calculator or computer. Copy and label them in Figure 1.

b. For positive constants b, the derivative of b^x is the limit of $\dfrac{b^{x+\Delta x} - b^x}{\Delta x}$ as $\Delta x \to 0$. Setting $x = 0$ and $\Delta x = 0.0001$ gives the approximation,

$$\left[\dfrac{d}{dx} b^x\right]_{x=0} \approx \dfrac{b^{0+0.0001} - b^0}{0.0001} = \dfrac{b^{0.0001} - 1}{0.0001}. \tag{1}$$

Use your calculator or computer to find this approximate derivative for $b = 2$ and $b = 3$.

c. Notice that the derivative of 2^x at $x = 0$ is less than 1, while the derivative of 3^x at $x = 0$ is greater than 1. There is a value of b between 2 and 3 such that the derivative of b^x at $x = 0$ is equal to 1. This value of b is denoted e. Find its approximate value with one decimal place accuracy by determining the value 2.1, 2.2, 2.3, 2.4, 2.5, 2.6, 2.7, 2.8, or 2.9 of b for which the approximate derivative (1) is closest to 1.

d. Find e to two decimal places. (You could continue the procedure to find the decimal value of e with whatever accuracy you want.)

e. The function e^x is probably given on your calculator or computer as $e \wedge x$ or $\exp(x)$. Generate its graph and add it to Figure 1.

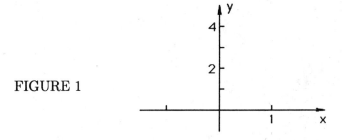

FIGURE 1

---

†In this worksheet numerical experiments are used to predict formulas for derivatives of exponential functions. See Calculator instructions 1B.3 for tips on working it with a calculator.

174 Graphing calculator workbook

Problem 2 Complete Table 1 of approximate decimal values e^x and approximate values of its derivative at $x = 0$, $x = 1$, $x = 2$ and $x = 3$. Use these results to predict a formula for the derivative of e^x as a function of x. If you cannot see an answer right away, calculate the approximate derivative $\dfrac{e^{x+\Delta x} - e^x}{\Delta x}$ for $x = 0, 1, 2$, and 3 with positive values of Δx that are smaller than 0.0001, or calculate e^x and the approximate derivative at other values of x.

x	$e^x \doteq$	$\dfrac{d}{dx} e^x \approx \dfrac{e^{x+0.0001} - e^x}{0.0001} \doteq$
0	1	1.000050000
1		
2		7.389425570
3	20.08553692	

TABLE 1

Problem 3 Complete Table 2 of approximate decimal values of 10^x and approximate values of its derivative at $x = 0$, $x = 1$, $x = 2$ and $x = 3$. Use these results and other calculations, if necessary, to predict a formula for the approximate derivative of 10^x as a function of x. The exact formula will be derived in your class.

x	$10^x \doteq$	$\dfrac{d}{dx} 10^x \approx \dfrac{10^{x+0.0001} - 10^x}{0.0001} \doteq$
0	1	2.302850210
1		
2		230.2850210
3	1000	

TABLE 2

Problem 4 Complete Table 3 of approximate decimal values of 2^x and approximate values of its derivative at $x = 0$, $x = 1$, $x = 2$ and $x = 3$. Use these results and other calculations, if necessary, to predict a formula for the approximate derivative of 2^x as a function of x. The exact formula will be derived in your class.

x	$2^x \doteq$	$\dfrac{d}{dx} 2^x \approx \dfrac{2^{x+0.0001} - 2^x}{0.0001} \doteq$
0	1	0.6931712000
1		
2	4	2.7726848200
3		

TABLE 3

Graphing calculator workbook

Worksheet 4A.2†

Derivatives of logarithms: Numerical experiments

Name _____ Date _____

Others in your group _____

Instructor, Teaching Assistant, and/or Recitation section _____

Directions Put first drafts of your calculations and answers on scratch paper. Take your time, work carefully, and discuss your solution with at least one other student before putting a final draft on this sheet or on other paper. Turn in all your work.

Problem 1a The logarithm to the base 10, $\log_{10} x$, is given as $\log x$ on most calculators and computers. Generate its graph for $-1 \leq x \leq 6, -2 \leq y \leq 2$ and copy it in Figure 1.

b. The logarithm to the base e, $\log_e x$, is probably given as $\ln x$ on your calculator or computer. Generate its graph with the range from part (a) and copy it in Figure 2.

c. Generate the graph of $\dfrac{\log_{10} x}{\ln x}$ for $-1 \leq x \leq 6, -0.2 \leq y \leq 1$ with y-scale $= 0.2$ and copy it in Figure 3. Then use it to explain how the graphs in Figures 1 and 2 are related.

Problem 2a Figure 4 shows the graph of $\ln x$ and its tangent lines at $x = 1, 2$ and 4. Use the approximation, $\dfrac{d}{dx} \ln x \approx \dfrac{\ln(x + 0.000001) - \ln x}{0.000001}$ and your calculator or computer to complete Table 1 with the approximate slopes of these lines and of the tangent line at $x = 10$.

x	$\dfrac{d}{dx} \ln x \approx \dfrac{\ln(x + 0.000001) - \ln x}{0.000001} \doteq$
1	0.999999500
2	
4	
10	

TABLE 1

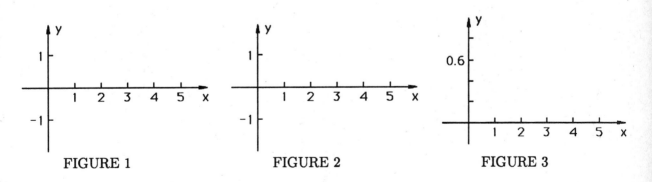

FIGURE 1   FIGURE 2   FIGURE 3

---

†Numerical experiments are used in this worksheet to predict formulas for derivatives of $\ln x$ and $\log_{10} x$. See Calculator instructions 1B.3 for tips on working it with a calculator.

b. Use the results in Table 1 to predict a formula for the derivative of $\ln x$. If you cannot guess the formula from these values, replace 0.000001 by smaller positive numbers in the expression for the approximate derivative or find the approximate derivatives at other values of x.

Problem 3a Figure 5 shows the graph of $\log x$ and its tangent lines at $x = 1, 2$ and 4. Use the approximation, $\dfrac{d}{dx} \log x \approx \dfrac{\log(x + 0.000001) - \log x}{0.000001}$ and your calculator or computer to complete Table 2 of approximate slopes of these lines and of the tangent lines at $x = 10$ and $x = 100$.

TABLE 2

x	$\dfrac{d}{dx} \log x \approx \dfrac{\log(x + 0.000001) - \log x}{0.000001} \doteq$
1	0.432942648
2	
4	
10	
100	

b. Use the results in Table 2 to predict an approximate formula for the derivative of $\log_{10} x$. If you cannot guess such a formula from these values, calculate the approximate derivative at other values of x.

Problem 4 How are the results of Problems 1c, 2b, and 3b related?

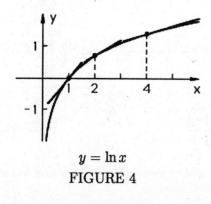

$y = \ln x$

FIGURE 4

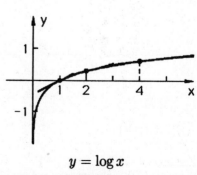

$y = \log x$

FIGURE 5

Graphing calculator workbook

Worksheet 4A.3†

The curious integrals of 1/x: Numerical experiments

Name _____ Date _____

Others in your group _____

Instructor, Teaching Assistant, and/or Recitation section _____

Directions Put first drafts of your calculations and answers on scratch paper. Take your time, work carefully, and discuss your solution with at least one other student before putting a final draft on this sheet or on other paper. Turn in all your work.

Problem 1a Use Riemann sum program 3B with $0 \leq x \leq 9, -0.2 \leq y \leq 1.2$ to find the Riemann sum approximation of $\int_1^2 \frac{1}{x}\,dx$ that is obtained by using three equal subintervals with the integrand evaluated at the left endpoints of the intervals ($A = 1, B = 2, N = 3, T = 0$). The program should generate a drawing like that in Figure 1.

b. Use the program with the same ranges to find the Riemann sum approximation of $\int_3^6 \frac{1}{x}\,dx$ with three equal subintervals and the integrand evaluated at the left endpoints ($A = 3, B = 6, N = 3, T = 0$). Copy the rectangles in Figure 2.

c. Next use the program with the same ranges to find the Riemann sum approximation of $\int_4^8 \frac{1}{x}\,dx$ with three equal subintervals and the integrand evaluated at the left endpoints ($A = 4, B = 8, N = 3, T = 0$). Copy the rectangles in Figure 3.

d. Why are the Riemann sums in Problems 1, 2 and 3 equal? Compare the widths and heights of the rectangles.

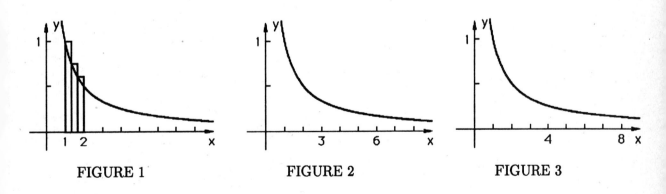

FIGURE 1   FIGURE 2   FIGURE 3

---

†Riemann sums are used in this worksheet to show that the integral of $1/x$ from 1 to b acts like a logarithm of b for $b > 0$. (In fact, the integral equals $\ln b$.) See Calculator instructions 1B.3 for tips on solving the problems with a calculator.

178   Graphing calculator workbook

    e.   Rework parts (a), (b), and (c) with one or more large values of N in place of 3. What is the result and why? What does this tell you about the three integrals

$$\int_1^2 \frac{1}{x}\, dx, \quad \int_2^4 \frac{1}{x}\, dx, \quad \text{and} \quad \int_3^6 \frac{1}{x}\, dx?$$

Problem 2   Use the result of Problem 1e to predict a formula relating $\displaystyle\int_1^b \frac{1}{x}\, dx$ and $\displaystyle\int_a^{ab} \frac{1}{x}\, dx$ for arbitrary positive a and b. Check it by making Riemann sum approximations of both integrals—with the same N and T—for various positive values of a and b.

Problem 3   Notice that $\displaystyle\int_1^{ab} \frac{1}{x}\, dx = \int_1^a \frac{1}{x}\, dx + \int_a^{ab} \frac{1}{x}\, dx$ for positive a and b.
Use this equation and the result of Problem 2 to predict a formula relating $\displaystyle\int_1^{ab} \frac{1}{x}\, dx, \int_1^a \frac{1}{x}\, dx$ and $\displaystyle\int_1^b \frac{1}{x}\, dx$.

Problem 4   The formula from Problem 3 shows that $\displaystyle L(a) = \int_1^a \frac{1}{x}\, dx$ has a property of logarithms. Explain.

Problem 5   Why does the function $L(x)$ also have the property $L(1) = 0$ of logarithms?

Problem 6   Use the formula from Problem 4 to show that $L(x)$ also satisfies

$$L(a^2) = 2L(a), \quad L(a^3) = 3L(a), \quad \text{and} \quad L(a^4) = 4L(a)$$

for all positive numbers a.

Worksheet 4B.1[†]

Applications of exponential functions

Name _____ Date _____

Others in your group _____

Instructor, Teaching Assistant, and/or Recitation section _____

Directions Put first drafts of your calculations and answers on scratch paper. Take your time, work carefully, and discuss your solution with at least one other student before putting a final draft on this sheet or on other paper. Turn in all your work.

Problem 1 Which of Figures 1 through 6 might be the graph of each of the following functions? Justify your conclusions.

a. A potato at room temperature is dropped into a pan of boiling water. $y(x)$ is the temperature (degrees Celsius) of the potato x minutes later. Water boils at 100° Celsius.

b. $y(x)$ is the size of a population as a function of the time x (years). The population is 100 at $x = 0$ and doubles every 10 years.

c. A saturated salt solution is at an equilibrium concentration of 120 parts per liter up to time $x = 0$, when the temperature is suddenly lowered, the equilibrium concentration changes to 100 parts per liter, and the salt begins to precipitate from the solution. $y(x)$ is the concentration of salt for $x \geq 0$.

d. A light is shining down on the ocean. $y(x)$ is its intensity (measured in candles) as a function of the depth x (meters) below the surface. The intensity is 100 units at $x = 0$.

e. The graph of $y(x)$ is a bell curve that describes the distribution of annual incomes x in a population, where x is measured in thousands of dolars. The area between the entire curve and the x-axis is 1 and the area between the curve and the x-axis for $a \leq x \leq b$ is the fraction of the population with incomes between a and b. The average income is 15 thousand dollars, and the maximum value of $y(x)$ is approximately 0.08.

f. $y(x)$ is the mass (grams) of a patch of moss at time x (years). The amount of moss varies because of changing climatic conditions, which repeat every year. The minimum amount of moss is 1 gram.

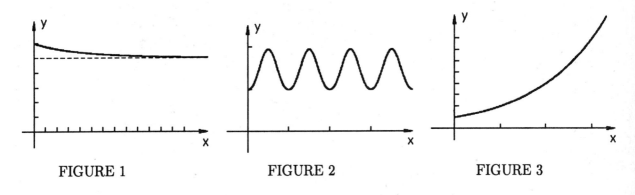

FIGURE 1   FIGURE 2   FIGURE 3

[†]This worksheet calls for matching descriptions of functions with their graphs and formulas. It does not use any calculus.

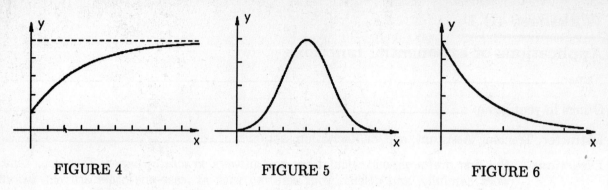

FIGURE 4   FIGURE 5   FIGURE 6

Problem 2   Use properties of the exponential function to pick formulas for the functions whose graphs are shown in Figures 1 through 6 from the list below. Give your reasoning and check your conclusions by generating the graphs on your calculator or computer. How far apart are the ticmarks on the x-axis in Figures 2 and 5? How far apart are the tickmarks on the y-axes in Figures 1, 3, 4, and 6 and why?

a. $y = \dfrac{1}{\sqrt{(50\pi)}} e^{-(x-15)^2/50}$ Use $(1/\sqrt{(50\pi)}) \, e \wedge (-(x-15)^2/50)$.

b. $y = 100(2^{x/10})$

c. $y = \dfrac{600}{6 - e^{-x/5}}$

d. $y = e^{(1-\cos(2\pi x))/3}$ Use $e \wedge ((1 - \cos(2\pi x))/3)$.

e. $y = 100 - 80e^{-x/3}$

f. $y = 100e^{-1.4x}$

Graphing calculator workbook   181

Worksheet 4B.2†

Applications of logarithms

Name _____ Date _____

Others in your group _____

Instructor, Teaching Assistant, and/or Recitation section _____

Directions   Put first drafts of your calculations and answers on scratch paper. Take your time, work carefully, and discuss your solution with at least one other student before putting a final draft on this sheet or on other paper. Turn in all your work.

Problem 1   Which of Figures 1 through 4 might be the graph of each of the following functions? Justify your conclusions.

   a. $y(x)$ is the magnitude of a star whose brightness is x, measured in a scale such that the star Pollux in the constellation Gemini has brightness 0.344 and magnitude 1.16; Vega in the constellation Lyra has brightness 1 and magnitude 0; while Sirius, the brightest star, has brightness 3.63 and magnitude -1.4. (When the English astronomer Pogson established this definition of the magnitude of stars in 1850, he chose to give dimmer stars greater magnitudes because since the time of the Greek astronomer Hipparchus (150 BC) the brightest stars had been referred to as "first magnitude stars" and less bright stars as "second" or "third magnitude stars.")

   b. $y(x)$ is the nonnegative solution of the equation $x = 10^{y^2}$.

   c. $y(x)$ is the magnitude, measured in the Richter scale, of an earthquake which produces x hundred thousand kilowatt-hours of energy. An earthquake that produces one hundred thousand kilowatt-hours of energy has magnitude 4.52, and an earthquake that produces five hundred thousand kilowatt-hours of energy has magnitude 4.99.

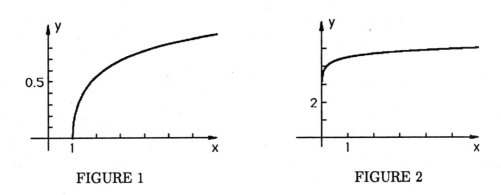

FIGURE 1   FIGURE 2

---

†This worksheet calls for matching descriptions of functions with their graphs and formulas. It does not use any calculus.

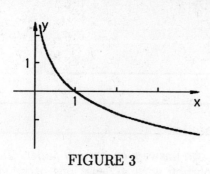
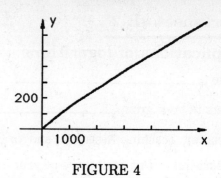

FIGURE 3 FIGURE 4

d. $y(x)$ is the approximate number of prime numbers $\leq x$ for large numbers x, as given by the "prime number theorem." According to that theorem there are approximately 145 prime numbers ≤ 1000 and 587 prime numbers ≤ 5000. (This is only an approximation. In fact, there are 168 prime numbers ≤ 1000 and 669 prime numbers ≤ 5000.)

Problem 2 Use properties or values of logarithms to pick formulas from the list below for the functions whose graphs are shown in Figures 1 through 4. Explain your reasoning. Then generate the graphs on your calculator or computer to check your choices.

a. $y = \dfrac{x}{\ln x}$

b. $y = -2.5 \log_{10} x$

c. $y = 0.67 \log_{10}(100000\, x) + 1.17$

d. $y = \sqrt{\log_{10} x}$

Graphing calculator workbook

Worksheet 4B.3†

Compound interest

Name _____ Date _____

Others in your group _____

Instructor, Teaching Assistant, and/or Recitation section _____

Directions Put first drafts of your calculations and answers on scratch paper. Take your time, work carefully, and discuss your solution with at least one other student before putting a final draft on this sheet or on other paper. Turn in all your work.

Problem 1 If P dollars are deposited in a savings account that pays $100R\%$ annual interest compounded N times a year and if no withdrawals or further deposits are made, then at times $t = j/N$ for positive integers j, when interest is paid, the balance is

$$B = P\left(1 + \frac{R}{N}\right)^{Nt} \quad \text{dollars.} \tag{1}$$

Between interest payments, as Nt increases from one integer to the next, the balance is constant. Consequently, the balance is given for all $t \geq 0$ by

$$B = P\left(1 + \frac{R}{N}\right)^{\text{Int}(Nt)} \quad \text{dollars} \tag{2}$$

where $\text{Int}(x)$ is the "integer part" of x, the greatest integer $\leq x$. For example, $\text{Int}(2.3) = 2$, $\text{Int}(\frac{1}{2}) = 0$, and $\text{Int}(\pi) = 3$.

If the interest is compounded continuously, then the balance for $t \geq 0$ is

$$B = Pe^{Rt} \quad \text{dollars.} \tag{3}$$

a. The horizontal lines in Figure 1 form the graph of a bank balance (as a function of t, measured in years) where the initial deposit is $\$100$ and the annual interest rate is 50% compounded annually. The continuous curve shows the balance with the

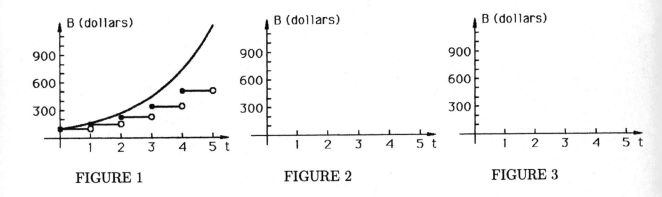

FIGURE 1   FIGURE 2   FIGURE 3

---

†This worksheet uses graphs of bank balances under compound interest to compare the different methods of compounding and to study equivalent rates for the different methods. See Calculator instructions 1B.3 for tips on working it with a calculator.

with the same initial deposit and interest rate if the interest is compounded continuously. Store the functions (2) and (3) in your calculator or computer with $P, R,$ and N as parameters. Then generate their graphs with $P = 100, R = 0.5,$ and $N = 1$ for $-0.1 \leq x \leq 5, -50 \leq y \leq 1000,$ and y-scale $= 100$. The graphs should look like the curve and five lines in Figure 1 without the dots and circles.

b. To compare 50% annual interest compounded semi-annually with 50% annual interest compounded continuously, change N to 2 and regenerate the graphs. Copy them in Figure 2.

c. Compare 50% annual interest compounded monthly to 50% annual interest compounded continuously by changing N to 12 and generating the graphs. Copy them in Figure 3.

d. Follow the instructions of part (c) for interest compounded daily by setting $N = 365$ and copy the graphs in Figure 4. Then compare the two balances at $t = 4$ by calculating (2) and (3) directly. Explain the results.

Problem 2 Figure 5 shows the graph of the balance from an initial deposit of $100 with 50% annual interest compounded annually, as in Figure 1, and with the equivalent rate compounded continuously (the rate that gives the same yield at the beginning of every year). Find this equivalent rate. As a partial check of your answer, generate the two graphs with this value for R and compare the result with Figure 5.

Problem 3 Figure 6 shows the graph of the balance in another account that had no withdrawals or other deposits after $t = 0$ (years). The scale on the B-axis is given in thousands of dollars.

a. What was the initial deposit? What was the method of compounding? What was the approximate annual interest rate?

b. Draw in Figure 6 the graph of the balance in an account with the same initial deposit and the equivalent interest rate compounded continuously. What is that approximate rate?

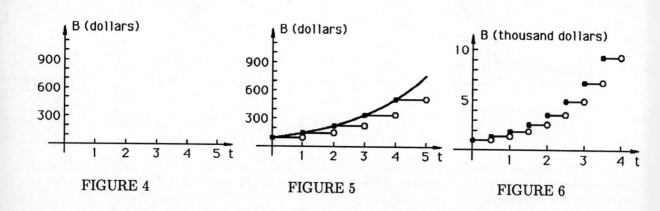

FIGURE 4　　　　　　　FIGURE 5　　　　　　　FIGURE 6

Graphing calculator workbook

Worksheet 4C.1†

Masterpiece around the corner

Name _____ Date _____

Others in your group _____

Instructor, Teaching Assistant, and/or Recitation section _____

Directions *Put first drafts of your calculations and answers on scratch paper. Take your time, work carefully, and discuss your solution with at least one other student before putting a final draft on this sheet or on other paper. Turn in all your work.*

Problem 1 Sonia wants to paint as large paintings as she can, but to get them out of her studio she has to slide them around a corner from a three-foot wide hall to a four-foot wide hall (Figure 1). Enter the following program in your calculator to illustrate a picture of width W feet being slid around the corner. If you have trouble entering any lines, look for similar lines in earlier programs for which the necessary key combinations are listed.

1.	Prgm9: HALLS	13.	: Line(10, 4, 10, 10)
2.	: All-Off	14.	: Pause
3.	: ClrDraw	15.	: Disp "WIDTH = "
4.	: Disp "HALLS"	16.	: Input W
5.	: Pause	17.	: $3 \to U$
6.	: $0 \to X$min	18.	: Lbl 1
7.	: $15 \to X$max	19.	: $\sqrt{(\,(10-U)^2+9\,)} \to D$
8.	: $0 \to Y$min	20.	: Line(U, 1, $U+W(10-U)/D$, $1+3W/D$)
9.	: $10 \to Y$max	21.	: $U+0.5 \to U$
10.	: Line(0, 1, 14, 1)	22.	: Pause
11.	: Line(14, 1, 14, 10)	23.	: If $U < 10$
12.	: Line(0, 4, 10, 4)	24.	: Goto 1

FIGURE 1

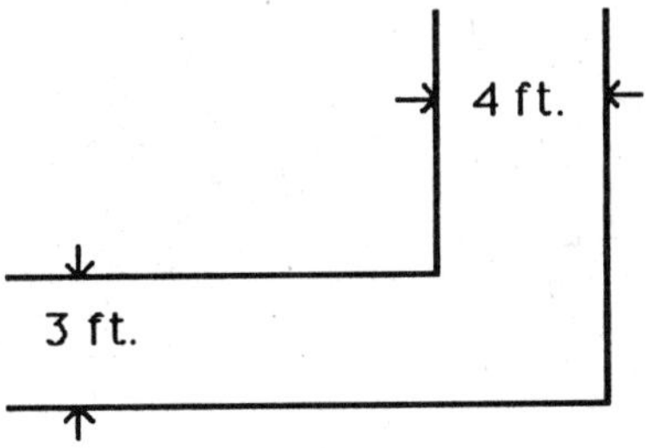

4 ft.

3 ft.

---
†This worksheet uses a calculator program to illustate a maximum/minimum problem that can be solved using inverse trigonometric functions. It is written for a Texas Instruments TI-81 calculator.

a. Experiment with the program, using different values of W, until you determine the approximate maximum width of a painting that can be slid around the corner. You can rerun the program with the ⌞ENTER⌟ key if you do no other operations after it has finished.

b. Use calculus to find the exact solution. Check your result with a graph on your calculator.

Graphing calculator workbook

Worksheet 4C.2†

What's the angle?

Name _____ Date _____

Others in your group _____

Instructor, Teaching Assistant, and/or Recitation section _____

Directions Put first drafts of your calculations and answers on scratch paper. Take your time, work carefully, and discuss your solution with at least one other student before putting a final draft on this sheet or on other paper. Turn in all your work.

Problem 1 Joe is driving east along a straight highway perpendicular to a thirty-mile long mountain range that begins fifty miles from the highway (Figure 1). He wants to photograph the mountain range from the point on the highway where it makes the largest horizontal angle at his camera, so it will fill as much of his picture as possible. Enter the following program in your calculator to show the angle for different positions on the road. If you have trouble entering any lines, look for similar lines in earlier programs for which the necessary key combinations are listed.

1. PrgmA: MTNVIEW
2. : All-Off
3. : ClrDraw
4. : Disp "MOUNTAIN VIEW"
5. : Pause
6. : $0 \to X$min
7. : $120 \to X$max
8. : $10 \to X$scale
9. : $0 \to Y$min
10. : $90 \to Y$max
11. : $10 \to Y$scale
12. : Line(0, 10, 120, 10)
13. : Pause
14. : Line(100, 50, 100, 80)
15. : Pause
16. : Lbl 1
17. : Disp "D = "
18. : Input D
19. : Line(100 − D, 10, 100, 50)
20. : Pause
21. : Line(100 − D, 10, 100, 80)
22. : Pause
23. : Disp "1 FOR MORE"
24. : Disp "0 TO QUIT"
25. : Input A
26. : If $A = 1$
27. : Goto 1

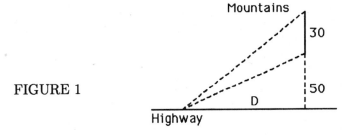

FIGURE 1

---

†This worksheet uses a calculator program to illustate a maximum/minimum problem that can be solved using inverse trigonometric functions. It is written for a Texas Instruments TI-81 calculator.

a. Experiment with the program, using different values of the distance D shown in Figure 1, until you determine the approximate value that yields the largest angle. You can rerun the program with the ENTER key if you do no other operations after it has finished.

b. Use calculus to find the exact value of the best choice of D. Then generate a graph to check your answer.

Graphing calculator workbook

Worksheet 4D.1†

Functions constructed from inverse trigonometric functions

Name _____ Date _____

Others in your group _____

Instructor, Teaching Assistant, and/or Recitation section _____

Directions Put first drafts of your calculations and answers on scratch paper. Take your time, work carefully, and discuss your solution with at least one other student before putting a final draft on this sheet or on other paper. Turn in all your work.

Problem 1 Figure 1 shows the graph of $\sin^{-1} x$ and of a function $f(x)$ that is formed by composing the three functions $x^2, 2x$, and $\sin^{-1} x$ in one of the six possible orders. What function is it? Check your conclusion by generating the graphs on your calculator or computer and by finding the coordinates of the dots in Figure 1. Then explain how the formula for $f(x)$ determines the shape of its graph.

Problem 2 Figure 2 shows the graph of a linear combination $g(x)$ of $\sin^{-1} x$, $\cos^{-1} x$, and $\tan^{-1} x$. What function is it? Check your conclusion by generating the graph on your calculator or computer. Then explain how the formula for $g(x)$ determines the shape of the graph.

Problem 3 Figure 3 shows the graphs of four functions given by $\tan^{-1}(Ax + B)$ with integer values of the constants A and B. What are the four functions? Check your conclusion by generating the graphs on your calculator or computer for $-6 \leq x \leq 6$, $-1.75 \leq y \leq 1.75$. Then explain how the formulas determine the shapes of the graph.

Problem 4 Generate the graph of $\sin^{-1}(\sin x)$ with $-2 \leq x \leq 11$, y-scale $= 1.57 \, (\doteq \pi/2)$, $-3 \leq y \leq 5.667$, and y-scale $= 1.57$ on your calculator or computer and copy it in Figure 4. Then explain its shape.

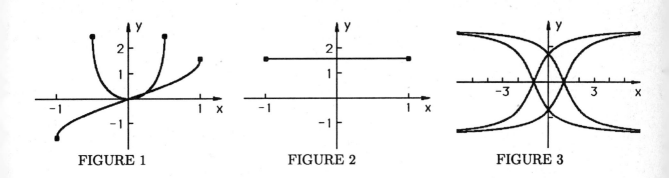

FIGURE 1   FIGURE 2   FIGURE 3

---

†This worksheet reviews properties of inverse trigonometric functions and their graphs. It does not use any calculus.

Problem 5 Generate the graph of $\cos^{-1}(\cos x)$ with the x- and y-ranges of Problem 4 and copy it in Figure 5. Explain its shape.

Problem 6 Generate the graph of $\cos(\sin^{-1} x)$ with the x- and y-ranges of Problem 4 and copy it in Figure 6. Explain its shape.

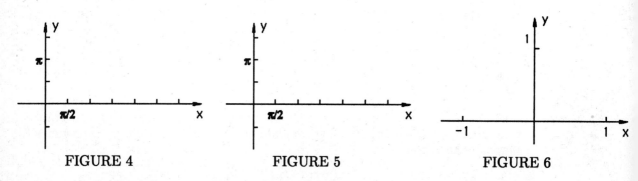

FIGURE 4   FIGURE 5   FIGURE 6

Problem 7 Generate the graph of $\tan(\tan^{-1} x)$ for $-6 \leq x \leq 6, -4 \leq y \leq 4$ on your calculator or computer and copy it in Figure 7. Explain its shape.

Problem 8 Generate the graph of $\tan^{-1}(\tan x)$ for $-2 \leq x \leq 11, -3 \leq y \leq 5.667$ on your calculator or computer and copy it in Figure 8. Explain its shape. (The graph on your calculator or computer might include extraneous vertical lines, which can be eliminated on some calculators by using their "dot" modes.)

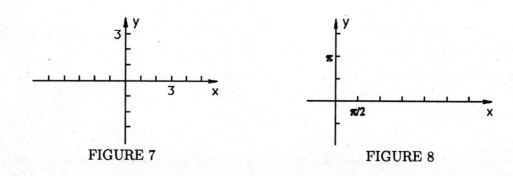

FIGURE 7   FIGURE 8

Graphing calculator workbook   191

Worksheet 4E.1[†]

Integrating rational functions: Initial investigations

Name _____ Date _____

Others in your group _____

Instructor, Teaching Assistant, and/or Recitation section _____

Directions   *Put first drafts of your calculations and answers on scratch paper. Take your time, work carefully, and discuss your solution with at least one other student before putting a final draft on this sheet or on other paper. Turn in all your work.*

Problem 1   The curve in Figure 1 is the graph of the sum $f(x)$ of two of the following functions:

$$\frac{3}{x-2}, \qquad \frac{2}{x-1}, \qquad \frac{1}{x^2}, \qquad 20x-10, \qquad \frac{5}{x^2-1}.$$

a.   Use the shape of the graph to determine which two functions are used. Justify your answer and check it by generating the graph on your calculator or computer with $-2 \le x \le 3, -20 \le y \le 20$, and y-scale $= 5$.

b.   Which of the following equals $f(x)$?

$$\frac{x^3-x^2+1}{x^2}, \qquad \frac{5x-7}{x^2-3x+2}, \qquad \frac{2x^2+x-1}{x^3-x^2}, \qquad \frac{3x^2+x-2}{x^3-2x^2}$$

c.   Use integration formulas to find the exact value of $\displaystyle\int_{1.1}^{2.5} f(x)\,dx$. Use the Riemann sum program 3B or Simpson's rule program 3F to check your answer.

FIGURE 1

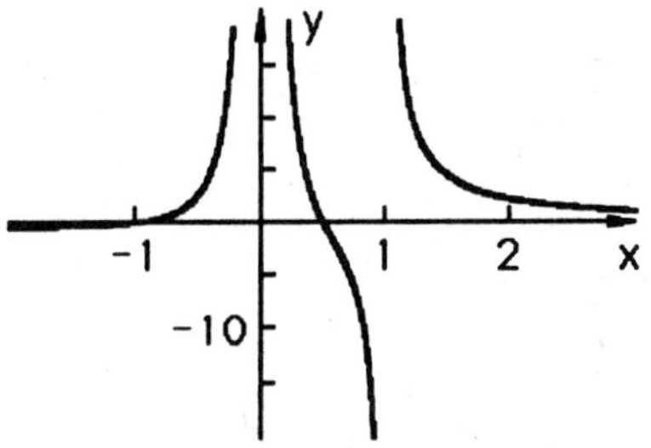

   [†]This worksheet is an introduction to the method of partial fractions and should be worked before you study that method.

Problem 2 Figure 2 shows the graph of

$$g(x) = Ax + \frac{B}{x^2+1}$$

for certain integers A and B. As indicated in the figure, the line $y = x$ is an asymptote of the graph.

a. Find A and B, and justify your choices. Check your answer by generating the graph on your calculator or computer with $-4.5 \leq x \leq 4.5$, $-3 \leq y \leq 3$.

b. Which of the following equals $g(x)$?

$$\frac{x^3+x-2}{x^2+1}, \quad \frac{x^2+3}{x^2+1}, \quad \frac{2x}{x^3+x}, \quad \frac{x-2}{x^2+1}$$

c. Use integration formulas to find the exact value of $\int_{-2}^{3} g(x)\,dx$. Then use Riemann sum program 3B or Simpson's rule program 3F to check your answer.

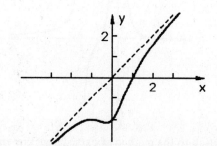

FIGURE 2

Graphing calculator workbook   193

Worksheet 5A.1[†]

Differential equations of growth and decay

Name _____ Date _____

Others in your group _____

Instructor, Teaching Assistant, and/or Recitation section _____

Directions *Put first drafts of your calculations and answers on scratch paper. Take your time, work carefully, and discuss your solution with at least one other student before putting a final draft on this sheet or on other paper. Turn in all your work.*

Problem 1 Each of problems A through D below is a growth or decay problem that leads to a differential equation for the function $y(t)$. Match the five problems to differential equations (i) through (iv).

Problems:

A The ice on a lake grows at a rate that is proportional to the reciprocal of its thickness. Find the thickness $y(t)$ of the ice as a function of the time t.

B A population grows at a rate proportional to its size. Find the population $y(t)$ as a function of the time t.

C A hot potato is taken out of the oven at time $t = 0$ into a kitchen that is at 20° Celsius. The rate of change of the potato's temperature is proportional to the difference between its temperature and that of the kitchen. Find the temperature $y(t)$ of the potato as a function of t.

D Find a function $y(t)$ whose rate of change with respect to t is $-2t$.

Differential equations:

(i) $\dfrac{dy}{dt} = 0.2y$ (ii) $\dfrac{dy}{dt} = \dfrac{20}{y}$

(iii) $\dfrac{dy}{dt} = -2t$ (iv) $\dfrac{dy}{dt} = 2(20 - y)$

Problem 2 A direction field for a differential equation is an array of short line segments as in Figures 1 through 4. If the differential equation is $\dfrac{dy}{dt} = f(t, y)$, then the slope of each line equals the value of $f(t, y)$ at the midpoint of that line. Match the direction fields to differential equations (i) through (iv) above. Explain how you make your choices and put the differential equations under the corresponding figures. (Hints: If the right side of the differential equation does not involve t, then the lines in each row are parallel; if the right side does not depend on y, then the lines in each column are parallel; and, in regions where the right side of the equation is positive [or negative, or large, or small], the lines have positive [negative, large, or small] slopes.)

_____

†This worksheet calls for matching problems that can be solved by differential equation to the differential equations, their direction fields, and solutions. It does not require any techniques for solving the differential equations.

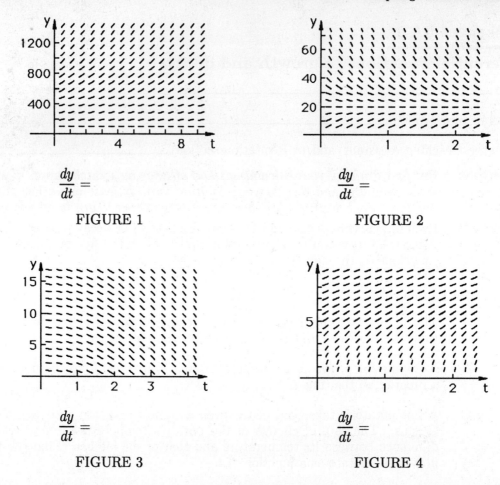

FIGURE 1

FIGURE 2

FIGURE 3

FIGURE 4

Problem 3a $y = 200e^{0.2t}$ is a solution of the differential equation of Figure 1. Generate its graph for $-1 \le t \le 10, -150 \le y \le 1500$ with y-scale $= 500$ and copy it in Figure 1. Then show that it is a solution of the corresponding differential equation.

b. $y = 20 + 40e^{-2t}$ is a solution of the differential equation of Figure 2. Generate its graph for $-0.25 \le t \le 2.5, -8 \le y \le 80$ with y-scale $= 10$ and copy it in Figure 2. Then show that it is a solution of the corresponding differential equation.

c. $y = 15 - t^2$ is a solution of the differential equation of Figure 3. Generate its graph for $-0.5 \le t \le 4, -2 \le y \le 18$ with y-scale $= 5$ and copy it in Figure 3. Then show that it is a solution of the corresponding differential equation.

d. $y = \sqrt{1 + 40t}$ is a solution of the differential equation of Figure 4. Generate its graph for $-0.25 \le t \le 2.5, -1 \le y \le 10$ with y-scale $= 5$ and copy it in Figure 4. Then show that it is a solution of the corresponding differential equation.

Graphing calculator workbook   195

Worksheet 5A.2[†]

Differential equations of velocity with resistance

Name _____ Date _____

Others in your group _____

Instructor, Teaching Assistant, and/or Recitation section _____

Directions   *Put first drafts of your calculations and answers on scratch paper. Take your time,*
work carefully, and discuss your solution with at least one other student before
putting a final draft on this sheet or on other paper. Turn in all your work.

Problem 1   Match differential equations I, II, and III to problems A, B, and C from which they
came. Justify your choices.

$$\textbf{(I)} \quad \frac{dv}{dt} = 9.6 \qquad \textbf{(II)} \quad \frac{dv}{dt} = \tfrac{1}{10}(96 - v) \qquad \textbf{(III)} \quad \frac{dv}{dt} = \tfrac{1}{5}(48 - v)$$

A. v is the downward velocity, measured in meters per second, of a ball that is falling
under the force of gravity but with no air resistance or other forces on it. The
time t is measured in seconds and the acceleration due to gravity is 9.6 meters per
second^2.

B. v is the downward velocity, measured in meters per second, of a suitcase that is
falling under the force of gravity with air resistance that is proportional to its
velocity. The suitcase's downward acceleration is zero when its downward velocity
is 48 meters per second. (This is called the "equilibrium" velocity because at that
velocity the upward force of air resistance equals the downward force of gravity. It
is also referred to as the "terminal" velocity because it is the limit of the suitcase's
velocity as $t \to \infty$.)

C. v is the downward velocity, measured in meters per second, of a rock that is falling
under the force of gravity with air resistance that is proportional to its velocity. Its
equilibrium velocity is 96 meters per second.

Problem 2   Match the differential equations in Problem 1 to the direction fields in Figures 1
through 3. Justify your choices and put the differential equations under the figures.

$$\frac{dv}{dt} = \qquad\qquad\qquad\qquad \frac{dv}{dt} = \qquad\qquad\qquad\qquad \frac{dv}{dt} =$$

FIGURE 1   FIGURE 2   FIGURE 3

_____

[†]This worksheet calls for matching problems that can be solved by differential equation to the differential equations, their direction fields, and solutions. It does not require any techniques for solving the differential
equations.

Problem 3 Generate the graphs of the solutions

(i) $v = 96 - 96e^{-t/10}$ (ii) $v = 9.6t$ (iii) $v = 48 - 48e^{-t/5}$

with your calculator or computer and copy them with the corresponding direction fields in Figures 1 through 3. Use x for t, y for v, $-3 \leq x \leq 30$, x-scale = 5, $-10 \leq y \leq 120$, and y-scale = 25. Then check that each function is a solution of the corresponding differential equation.

Problem 4 Match differential equations IV, V, and VI to problems D, E, and F from which they came. Justify your choices.

(IV) $\dfrac{dv}{dt} = 200 - 5v$ (V) $\dfrac{dv}{dt} = 45 + 45\cos(15t)$

(VI) $\dfrac{dv}{dt} = -\tfrac{1}{8}v^2$

D. v is the horizontal velocity, measured in miles per hour, of a model car whose acceleration, due to its faltering engine, oscillates between 0 and 90 miles per hour^2. There are no other forces on the car and time is measured in hours.

E. v is the horizontal velocity, measured in feet per minute, of a motor boat that has its engine turned off and is slowing down because of water and air resistance. The resistance is proportional to the square of the boat's velocity, and there are no other forces on it. Time is measured in minutes.

F. v is the horizontal velocity, measured in miles per hour, of a truck whose engine exerts a constant force and which is subject to air and rolling resistance that is proportional to its velocity. Time is measured in hours.

Problem 5 Match the differential equations in Problem 4 to the direction fields in Figures 4 through 6. Put the differential equations beneath the figures

Problem 6 (a) $v = 40(1 + 5t)^{-1}$ is a solution of the differential equation of Figure 4,
(b) $v = 45t + 3\sin(15t)$ is a solution of the differential equation of Figure 5, and
(c) $v = 40 - 40e^{-5t}$ is a solution of the differential equation of Figure 6. Generate their graphs and copy them with the direction fields, using $-0.1 \leq t \leq 1$, t-scale = 0.25, $-5 \leq v \leq 50$, and v-scale = 10. Then show that each function is a solution of the corresponding differential equation.

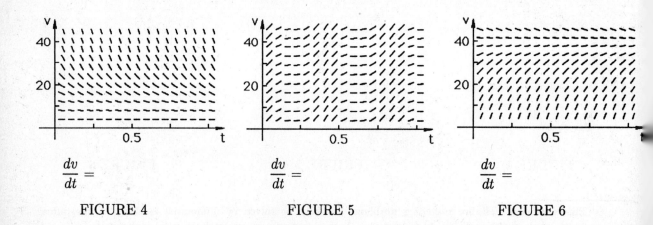

FIGURE 4   FIGURE 5   FIGURE 6

Worksheet 5A.3[†]

Differential equations of families of curves

Name _____ Date _____

Others in your group _____

Instructor, Teaching Assistant, and/or Recitation section _____

Directions Put first drafts of your calculations and answers on scratch paper. Take your time, work carefully, and discuss your solution with at least one other student before putting a final draft on this sheet or on other paper. Turn in all your work.

Problem 1 Match differential equations I, II, and III to their direction fields in Figures 1 through 3. Explain your choices and put the differential equations under the figures.

(I) $\dfrac{dy}{dx} = 3xy^{1/3}$ (II) $\dfrac{dy}{dx} = y \cos x$ (III) $\dfrac{dy}{dx} = \dfrac{1}{y^2}$

Problem 2a The family of functions $y = (x^2 + C)^{3/2}$, where C denotes an arbitrary positive constant, are solutions of one of the differential equations in Problem 1. Which is it? Show they are solutions. Then generate their graphs for $C = 1, 2, 3, 4$ with your calculator or computer and copy them in the figure with the corresponding direction field. Use the ranges given with the direction field.

b. The family of functions $y = Ce^{\sin x}$, with C an arbitrary constant, are solutions of another of the differential equations in Problem 1. Which is it? Show they are solutions. Then generate their graphs for $C = \pm 1, \pm 2$ and copy them with the corresponding direction field.

c. Show that the functions $y = (3x + C)^{1/3}$ with C an arbitrary constant are solutions of the remaining differential equation in Problem 1. Generate their graphs for $C = \pm 3, \pm 6$ and copy them with their direction field.

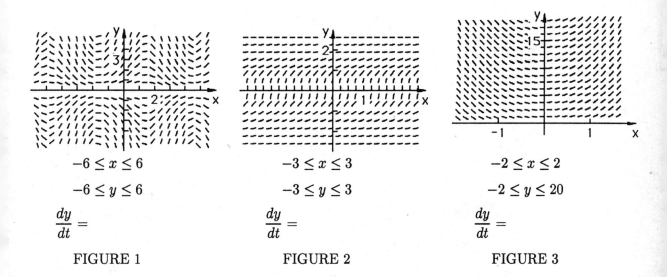

FIGURE 1 $-6 \le x \le 6$, $-6 \le y \le 6$, $\dfrac{dy}{dt} =$

FIGURE 2 $-3 \le x \le 3$, $-3 \le y \le 3$, $\dfrac{dy}{dt} =$

FIGURE 3 $-2 \le x \le 2$, $-2 \le y \le 20$, $\dfrac{dy}{dt} =$

[†]This worksheet calls for matching problems that can be solved by differential equation to the differential equations, their direction fields, and solutions. It does not require any techniques for solving the differential equations.

Problem 3 Figures 4 and 5 show members of the following four families of curves. Which are which?

(i) $y = \dfrac{C}{x}$

(ii) $y = \tfrac{1}{2}x^2 + C$

(iii) $y = -\ln|x| + C$

(iv) $y = \pm\sqrt{x^2 + C}$

Problem 4 Show that each family of curves in Problem 3 consists of solutions of one of the following differential equations.

A. $\dfrac{dy}{dx} = x$

B. $\dfrac{dy}{dx} = \dfrac{x}{y}$

C. $\dfrac{dy}{dx} = -\dfrac{y}{x}$

D. $\dfrac{dy}{dx} = -\dfrac{1}{x}$

Problem 5 The two families of curves in Figure 4 are called orthogonal trajectories because each curve intersects all of the curves in the other family at right angles (their tangent lines are perpendicular where the intersect). Use the results of Problem 4 to show this is the case.

Problem 6 Follow the instructions of Problem 5 for the two families of curves in Figure 5.

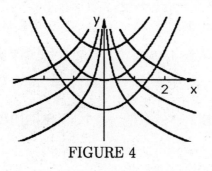

FIGURE 4

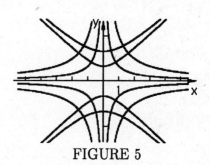

FIGURE 5

Graphing calculator workbook

Direction field program 5B

Texas Instruments TI-81

This program draws a direction field (slope field) of the differential equation $\dfrac{dv}{dx} = Y_1(x, v)$ with the function $Y_1(x, v)$ in the $Y =$ menu. The variable v is used instead of y because y cannot be used in the $Y =$ menu. The user chooses the number of columns of lines to be drawn. Subprogram ProgP allows for generating the graphs of $Y_2(x), Y_3(x)$, or $Y_4(x)$ after the direction field. It is given as a subprogram so it can be used with other programs.

Subprogram P for generating graphs

1.   Prgm P: PLOTYK
2.   : Lbl 1
3.   : Disp "∅ FOR NO PLOT"   (∅ denotes zero.)
4.   : Disp "K TO PLOT YK"
5.   : Input θ (θ is the Greek letter "theta.")
6.   : If $\theta = 1$
7.   : DrawF Y_1
8.   : If $\theta = 2$
9.   : DrawF Y_2
10.   : If $\theta = 3$
11.   : DrawF Y_3
12.   : If $\theta = 4$
13.   : DrawF Y_4
14.   : If $\theta \neq \emptyset$
15.   : Pause
16.   : If $\theta \neq \emptyset$
17.   : Goto 1
18.   : End

The main program

1.   Prgm5: DIRFIELD
2.   : All-Off
3.   : ClrDraw
4.   : Disp "USE "
5.   : Disp "$DV/DX = Y_1(X, V)$ "
6.   : Disp "HOW MANY COLUMNS"
7.   : Input N
8.   : $(X\max - X\min)/N \to D$
9.   : $3(Y\max - Y\min)/(2N) \to H$
10.   : $Y\min + H/2 \to V$
11.   : Lbl 1

12.	$: X\min + D/2 \to U$
13.	: Lbl 2
14.	$: U \to X$
15.	$: Y_1 \to M$
16.	$: D * \emptyset.35 \to W$
17.	: If abs $M > (H/D)$
18.	$: \emptyset.35 * H/\text{abs } M \to W$
19.	: Line$(U - W, V - MW, U + W, V + MW)$
20.	$: U + D \to U$
21.	: If $U < X\max$
22.	: Goto 2
23.	$: V + H \to V$
24.	: If $V < Y\max$
25.	: Goto 1
26.	: Pause
27.	: PrgmP

Entering the programs

Press 2nd QUIT to display the home screen. Then press PRGM for the menu of programs. Press ▶ so that EDIT is highlighted. Press ALPHA P for program #P and 5 for program #5.

Line-by-line instructions for entering the programs follow. To make a correction, move the cursor to the error and use INS and DEL as needed. To return to the program from other screens, press QUIT or follow the steps in the previous paragraph. If that does not work or you want to stop the running of the program, press ON 2 .

Subprogram P

1. Press P L O T Y K ENTER to identify the program.

---

2. Press PRGM 1 1 ENTER to start a loop.

---

3. Press PRGM ▶ 1 ALPHA " ∅ 2nd A-LOCK _ F O R _ N O _ P L O T " ENTER . ∅ denotes zero and _ is the space on the zero key.

Graphing calculator workbook   201

4. Press [PRGM] [▶] [1] [2nd] [A-LOCK] ["] [K] [_] [T] [O]
   [_] [P] [L] [O] [T] [_] [Y] [K] ["] [ENTER].

5. Press [PRGM] [▶] [2] [ALPHA] [θ] [ENTER] . We use θ here since it
   is not used in any of our main programs.

6. Press [PRGM] [3] [ALPHA] [θ] [2nd] [TEST] [1] [1] [ENTER].

7. Press [2nd] [DRAW] [6] [2nd] [Y-VARS] [1] [ENTER] .

8. Press [PRGM] [3] [ALPHA] [θ] [2nd] [TEST] [1] [2] [ENTER].

9. Press [2nd] [DRAW] [6] [2nd] [Y-VARS] [2] [ENTER] .

10. Press [PRGM] [3] [ALPHA] [θ] [2nd] [TEST] [1] [3] [ENTER].

11. Press [2nd] [DRAW] [6] [2nd] [Y-VARS] [3] [ENTER] .

12. Press [PRGM] [3] [ALPHA] [θ] [2nd] [TEST] [1] [4] [ENTER].

13. Press [2nd] [DRAW] [6] [2nd] [Y-VARS] [4] [ENTER] .

14. Press [PRGM] [3] [ALPHA] [θ] [2nd] [TEST] [2] [∅] [ENTER]
    .

15. Press [PRGM] [6] [ENTER] .

16. Press [PRGM] [3] [ALPHA] [θ] [2nd] [TEST] [2] [∅] [ENTER].

17. Press [PRGM] [2] [1] [ENTER] .

18. Press [PRGM] [7] [ENTER] . Execution returns to the main program.

202   Graphing calculator workbook

The main program

1. Press D I R F I E L D to identify the program.

2. Press 2nd Y-VARS ◄ 1 ENTER to unselect the functions in the $Y =$ menu.

3. Press 2nd DRAW 1 ENTER to clear the graphics screen.

4. Press PRGM ► 1 2nd A-LOCK " U S E " ENTER .

5. Press PRGM ► 1 2nd A-LOCK " D V ALPHA ÷ 2nd A-LOCK D X|T 2nd TEST 1 2nd Y-VARS 1 (2nd A-LOCK X|T , V ALPHA) ALPHA " ENTER .

6. Press PRGM ► 1 2nd A-LOCK " H O W - M A N Y - C O L U M N S " ENTER .

7. Press PRGM ► 2 ALPHA N ENTER .

8. Press (VARS ◄ 2 − VARS ◄ 1) ÷ ALPHA N STO► D ENTER . Direction lines will be drawn in N columns of squares. This is the width of the squares, measured in x-units.

9. Press 3 (VARS ◄ 5 − VARS ◄ 4) ÷ (2 ALPHA N) STO► H ENTER . This is the height of the squares, measured in y-units.

10. Press VARS ◄ 4 + ALPHA H ÷ 2 STO► V ENTER . This sets V equal to the v-coordinate of the middle of the first row of squares.

11. Press PRGM 1 1 ENTER for the beginning of a loop that is executed for each row of squares.

Graphing calculator workbook   203

---

12. Press $\boxed{\text{VARS}}$ $\boxed{\blacktriangleleft}$ $\boxed{1}$ $\boxed{+}$ $\boxed{\text{ALPHA}}$ $\boxed{D}$ $\boxed{\div}$ $\boxed{2}$ $\boxed{\text{STO}\blacktriangleright}$ $\boxed{U}$
$\boxed{\text{ENTER}}$. This sets U equal to the u-coordinate of the middle of the first column of squares. We have to use U instead of X here because X changes to the x-coordinate of the pixel being drawn when points are plotted.

---

13. Press $\boxed{\text{PRGM}}$ $\boxed{1}$ $\boxed{2}$ $\boxed{\text{ENTER}}$. This is the beginning of a loop that is executed for each square in the row.

---

14. Press $\boxed{\text{ALPHA}}$ $\boxed{U}$ $\boxed{\text{STO}\blacktriangleright}$ $\boxed{X|T}$ $\boxed{\text{ENTER}}$ to set X equal to U for evaluating Y_1.

---

15. Press $\boxed{\text{2nd}}$ $\boxed{\text{Y-VARS}}$ $\boxed{1}$ $\boxed{\text{STO}\blacktriangleright}$ $\boxed{M}$ $\boxed{\text{ENTER}}$ to store the slope of the direction field at (X, V) as M.

---

16. Press $\boxed{\text{ALPHA}}$ $\boxed{D}$ $\boxed{\times}$ $\boxed{0}$ $\boxed{.}$ $\boxed{3}$ $\boxed{5}$ $\boxed{\text{STO}\blacktriangleright}$ $\boxed{W}$ $\boxed{\text{ENTER}}$.
This is half the width of the slope line to be drawn if it extends from the left to the right side of its square.

---

17. Press $\boxed{\text{PRGM}}$ $\boxed{3}$ $\boxed{\text{2nd}}$ $\boxed{\text{ABS}}$ $\boxed{\text{ALPHA}}$ $\boxed{M}$ $\boxed{\text{2nd}}$ $\boxed{\text{TEST}}$ $\boxed{3}$
$\boxed{(}$ $\boxed{\text{ALPHA}}$ $\boxed{H}$ $\boxed{\div}$ $\boxed{\text{ALPHA}}$ $\boxed{D}$ $\boxed{)}$ $\boxed{\text{ENTER}}$.
If $|M| > \dfrac{H}{D}$, the next line is executed. This makes the slope line narrower if it is steep so it does not extend above and below its square.

---

18. Press $\boxed{0}$ $\boxed{.}$ $\boxed{3}$ $\boxed{5}$ $\boxed{\times}$ $\boxed{\text{ALPHA}}$ $\boxed{H}$ $\boxed{\div}$ $\boxed{\text{2nd}}$ $\boxed{\text{ABS}}$
$\boxed{\text{ALPHA}}$ $\boxed{M}$ $\boxed{\text{STO}\blacktriangleright}$ $\boxed{W}$ $\boxed{\text{ENTER}}$.

---

19. Press $\boxed{\text{2nd}}$ $\boxed{\text{DRAW}}$ $\boxed{2}$ $\boxed{\text{ALPHA}}$ $\boxed{U}$ $\boxed{-}$ $\boxed{\text{ALPHA}}$ $\boxed{W}$ $\boxed{\text{ALPHA}}$
$\boxed{,}$ $\boxed{ALPHA}$ $\boxed{V}$ $\boxed{-}$ $\boxed{\text{ALPHA}}$ $\boxed{M}$ $\boxed{\text{ALPHA}}$ $\boxed{W}$ $\boxed{\text{ALPHA}}$
$\boxed{,}$ $\boxed{\text{ALPHA}}$ $\boxed{U}$ $\boxed{+}$ $\boxed{\text{ALPHA}}$ $\boxed{W}$ $\boxed{\text{ALPHA}}$ $\boxed{,}$ $\boxed{ALPHA}$
$\boxed{V}$ $\boxed{+}$ $\boxed{\text{ALPHA}}$ $\boxed{M}$ $\boxed{\text{ALPHA}}$ $\boxed{W}$ $\boxed{)}$ $\boxed{\text{ENTER}}$ to draw the
slope line.

---

20. Press $\boxed{\text{ALPHA}}$ $\boxed{U}$ $\boxed{+}$ $\boxed{\text{ALPHA}}$ $\boxed{D}$ $\boxed{\text{STO}\blacktriangleright}$ $\boxed{U}$ $\boxed{\text{ENTER}}$.
U is increased for the next square.

---

204   Graphing calculator workbook

---

21. Press $\boxed{\text{PRGM}}$ $\boxed{3}$ $\boxed{\text{ALPHA}}$ $\boxed{U}$ $\boxed{\text{2nd}}$ $\boxed{\text{TEST}}$ $\boxed{5}$ $\boxed{\text{VARS}}$ $\boxed{\blacktriangleleft}$

$\boxed{2}$ $\boxed{\text{ENTER}}$. If the last slope line in the row has not yet been drawn, then U is less than X max and the next command is executed to repeat this inner loop.

---

22. Press $\boxed{\text{PRGM}}$ $\boxed{2}$ $\boxed{2}$ $\boxed{\text{ENTER}}$.

---

23. Press $\boxed{\text{ALPHA}}$ $\boxed{V}$ $\boxed{+}$ $\boxed{\text{ALPHA}}$ $\boxed{H}$ $\boxed{\text{STO}\blacktriangleright}$ $\boxed{V}$ $\boxed{\text{ENTER}}$.
V is increased for the next row of slope lines.

---

24. Press $\boxed{\text{PRGM}}$ $\boxed{3}$ $\boxed{\text{ALPHA}}$ $\boxed{V}$ $\boxed{\text{2nd}}$ $\boxed{\text{TEST}}$ $\boxed{5}$ $\boxed{\text{VARS}}$ $\boxed{\blacktriangleleft}$

$\boxed{5}$ $\boxed{\text{ENTER}}$. If the top row of slope lines has not yet been drawn, the next command is executed to repeat the outer loop.

---

25. Press $\boxed{\text{PRGM}}$ $\boxed{2}$ $\boxed{1}$ $\boxed{\text{ENTER}}$.

---

26. Press $\boxed{\text{PRGM}}$ $\boxed{6}$ $\boxed{\text{ENTER}}$ to have the program pause and display the screen until the user presses $\boxed{\text{ENTER}}$.

---

27. Press $\boxed{\text{PRGM}}$ $\boxed{\blacktriangleleft}$ $\boxed{\text{ALPHA}}$ $\boxed{P}$ $\boxed{\text{ENTER}}$ to run subprogram P which allows for generating the graphs of $Y_2(X), Y_3(X)$, or $Y_4(X)$ with the direction field.

Using the program

Before running the program, select the ranges of x and y for the graph and $f(x,y)$ as Y_1 in the $Y=$menu with v in place of y in the differential equation.

Example   Run the direction field program to generate the direction field of $\dfrac{dy}{dx} = \frac{1}{4}y^2 \sin x$ with 14 columns of slope lines for $-1 \leq x \leq 11, -0.5 \leq y \leq 5$. Then put on the same screen the graph of the solution $y = \dfrac{4}{\cos x + 2}$ with the initial condition $y(0) = \frac{4}{3}$.

Solution   Press $\boxed{\text{RANGE}}$ and enter -1 for x-min, 11 for x-max, 1 for x-scale, -0.5 for y-min, 5 for y-max, and 1 for y-scale. Set $Y_1 = 0.25V^2 \sin X$ for the direction field (using V in place of y) and $Y_2 = 4/(\cos X + 2)$ for the graph of the solution.

Press $\boxed{\text{PRGM}}$ $\boxed{7}$ $\boxed{\text{ENTER}}$. The statement "USE $DV/DX = Y_1(X,V)$" appears to remind you that the function $f(x,y)$ in the differential equation should be given as Y_1 in the $Y=$menu with V in place of y. After the prompt "HOW MANY COLUMNS?" enter 14 and press $\boxed{\text{ENTER}}$. The direction field should be drawn as in Figure 1, but with fewer slope lines. Press $\boxed{\text{ENTER}}$. The prompt

Graphing calculator workbook

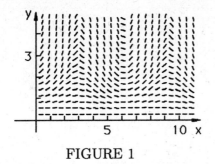
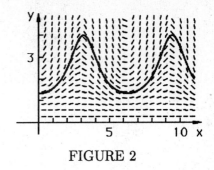

FIGURE 1   FIGURE 2

"0 FOR NO PLOT, K TO PLOT YK" should appear. Press 2 ENTER to add the graph of the solution as in Figure 2.

Graphing calculator workbook

Graphing calculator workbook

Worksheet 5C.1[†]

Euler's method

Name _____ Date _____

Others in your group _____

Instructor, Teaching Assistant, and/or Recitation section _____

Directions   *Put first drafts of your calculations and answers on scratch paper. Take your time, work carefully, and discuss your solution with at least one other student before putting a final draft on this sheet or on other paper. Turn in all your work.*

Problem 1   Figure 1 shows the direction field of the differential equation

$$\frac{dv}{dx} = \tfrac{1}{4} v^2 \sin x \tag{1}$$

and the graph of an approximate solution obtained for $0 \leq x \leq 5$ by Euler's method with five subintervals and the initial condition

$$v(0) = \tfrac{4}{3}. \tag{2}$$

With Euler's method, the graph of the approximate solution consists of one line segment for each subinterval, and the slope of each of the segments is equal to the value at its left end of the function given in the differential equation—which in this case is $\tfrac{1}{4} v^2 \sin x$. Consequently, each segment is parallel to the line in the direction field at its left end. The left end of the first segment is placed at

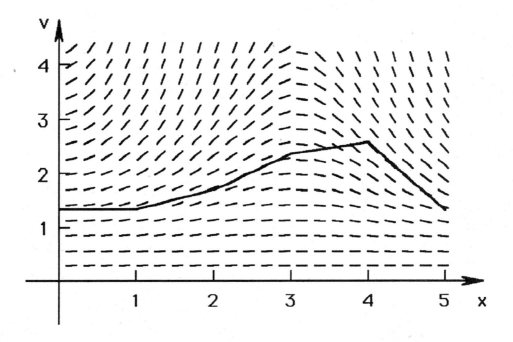

FIGURE 1

[†] This worksheet contains all of the information about Euler's method that it requires.

$x = 0, v = 4/3 \doteq 1.333333$ so that the approximate solution satisfies the initial condition (2). The left end of each of the other four segments is placed at the right end of the previous segment to make the resulting function be continuous.

Table 1 shows the slopes and coordinates of the ends of the first two segments with six-decimal place accuracy. Fill in the table for the last three segments.[†] You you should obtain the value at $x = 5$ given in the table, except perhaps for a round-off error in the last digit.

	Left end		Slope $= \frac{1}{4} v^2 \sin x$	Right end	
1	$x = 0$	$v \doteq 1.333333$	0	$x = 1$	$v \doteq 1.333333$
2	$x = 1$	$v \doteq 1.333333$	0.373987	$x = 2$	$v \doteq 1.707320$
3	$x = 2$	$v \doteq$		$x = 3$	$v \doteq$
4	$x = 3$	$v \doteq$		$x = 4$	$v \doteq$
5	$x = 4$	$v \doteq$		$x = 5$	$v \doteq 1.320296$

TABLE 1

Problem 2 Fill in Table 2 with the calculations for Euler's approximate solution of (1) and (2) for $0 \le x \le 5$ with ten subintervals. Draw the graph of this approximate solution in Figure 1. Then generate the graph of the exact solution $v = 4/(2 + \cos x)$ of (1)–(2) for $0.5 \le x \le 5, -0.5 \le v \le 4.5$ and copy it also in Figure 1. The graphs illustrate how well the two functions obtained by Euler's method approximate the exact solution. What happens? What would you expect to get with more subintervals?

	Left end		Slope $= \frac{1}{4} v^2 \sin x$	Right end	
1	$x = 0.0$	$v \doteq 1.333333$		$x = 0.5$	$v \doteq$
2	$x = 0.5$	$v \doteq$	0.213078	$x = 1.0$	$v \doteq 1.439872$
3	$x = 1.0$	$v \doteq$		$x = 1.5$	$v \doteq 1.657943$
4	$x = 1.5$	$v \doteq$		$x = 2.0$	$v \doteq$
5	$x = 2.0$	$v \doteq$		$x = 2.5$	$v \doteq 2.455637$
6	$x = 2.5$	$v \doteq$		$x = 3.0$	$v \doteq$
7	$x = 3.0$	$v \doteq$		$x = 3.5$	$v \doteq$
8	$x = 3.5$	$v \doteq$		$x = 4.0$	$v \doteq$
9	$x = 4.0$	$v \doteq$		$x = 4.5$	$v \doteq$
10	$x = 4.5$	$v \doteq$		$x = 5.0$	$v \doteq 1.502944$

TABLE 2

[†] See Calculator instructions 1B.3 for tips on solving this problem with a calculator.

Runge-Kutta program 5D[†]

Texas Instruments TI-81

This program uses a Runge-Kutta method with N subintervals to generate the graph of an approximate solution of $\dfrac{dv}{dx} = Y_1(x,v)$ in the interval $A \le x \le B$ with initial condition $v(A) = v_0$ and to display the value of the approximation at $x = B$. The variable v is used in place of y because y cannot be used in the Y= menu. The interval $A \le x \le B$ is partitioned into N equal subintervals of length $D = (B - A)/N$, and the graph of the approximate solution is constructed from line segments over the N subintervals. (See Figure 1 for the case of $N = 5$.) The left end of the first line segment (on the left) is put at (A, v_0) so the approximation satisfies the initial condition $v(A) = v_0$.

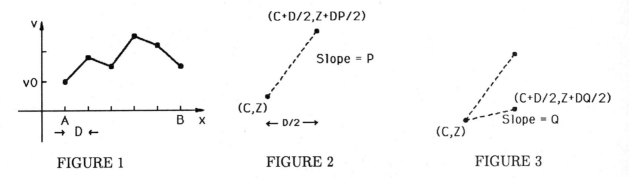

FIGURE 1   FIGURE 2   FIGURE 3

Euler's method, which is discussed in Worksheet 5C.1, uses similar line segments with the slope of each line segment taken to be the value of $Y_1(x,v)$ at its left endpoint. The Runge-Kutta method, which is generally much more accurate, takes a weighted average

(1) $$M = (P + 2Q + 2R + S)/6$$

of values of $Y_1(x,v)$ at four different points as the slope of the line segment.

In the program given below, (C, Z) denotes the coordinates of the left end of the line segment that is being constructed. Initially C is set equal to A and Z is set equal to v_0. The value of $Y_1(x,v)$ at (C, Z) is calculated and stored as P. A line of this slope determines the point $(C + D/2, Z + DP/2)$ (Figure 2), and the value of $Y_1(x,v)$ at this point is stored as Q. This slope and the point (C, Z) are used to determine the point $(C + D/2, Z + DQ/2)$ (Figure 3). The value of $Y_1(x,v)$ at this point is stored as R, and this slope and the point (C, Z) are used to determine the point $(C + D, Z + DR)$ (Figure 4). Finally, the value of $Y_1(x,v)$ at this point is stored as S.

The weighted average (1) of the values P, Q, R, and S of $Y_1(x,v)$ at the four points in Figure 4 is calculated and stored as M. This slope is used with the point (C, Z) to determine the point $(T, V) = (C + D, Z + DM)$, which is taken as the right endpoint of the line segment in the graph of the approximate solution (Figure 5). If this is not the last segment of the graph, then T is less than B, C is set equal to T, Z is set equal to V, and the next segment is drawn.

The screen should be cleared with the keys 2nd   QUIT   2nd   DRAW   1   ENTER   before running the program unless you want the graph of the approximate solution to be shown with a previously generated picture. Subprogram P, which is listed with the direction field program 5B, is included so the graph of Y_2, Y_3 or Y_4 can be added.

[†]This worksheet includes an explanation of the Runge-Kutta method that it uses.

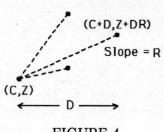
FIGURE 4

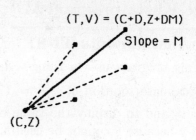
FIGURE 5

The program (244 bytes)

1.   Prgm 6: RUNGE
2.   : All-Off
3.   : Disp "USE"
4.   : Disp "$DV/DX = Y_1(X, V)$"
5.   : Disp "$A =$"
6.   : Input A
7.   : Disp "$B =$"
8.   : Input B
9.   : Disp "$V\emptyset =$" (Note: $\emptyset$ denotes zero.)
10.   : Input V
11.   : Disp "$N =$"
12.   : Input N
13.   : $(B - A)/N \to D$
14.   : $A \to T$
15.   : Lbl 1
16.   : $T \to C$
17.   : $V \to Z$
18.   : $C \to X$
19.   : $Y_1 \to P$
20.   : $C + D/2 \to X$
21.   : $Z + DP/2 \to V$
22.   : $Y_1 \to Q$
23.   : $Z + DQ/2 \to V$
24.   : $Y_1 \to R$
25.   : $C + D \to X$
26.   : $Z + DR \to V$
27.   : $Y_1 \to S$
28.   : $(P + 2Q + 2R + S)/6 \to M$
29.   : $X \to T$
30.   : $Z + DM \to V$

Graphing calculator workbook   211

31.   : Line(C, Z, T, V)
32.   : If $T < B$
33.   : Goto 1
34.   : Pause
35.   : Prgm P
36.   : If $\theta \neq \emptyset$ (Note: θ is theta and $\emptyset$ is zero.)
37.   : Pause
38.   : Disp "THE METHOD GIVES"
39.   : Disp "V = "
40.   : Disp V
41.   : Disp "AT X = "
42.   : Disp B

Entering the programs

Press $\boxed{\text{2nd}}$ $\boxed{\text{QUIT}}$ to display the home screen and $\boxed{\text{PRGM}}$ and $\boxed{\blacktriangleright}$ so that EDIT is highlighted and $\boxed{6}$ or another number if program 6 has already been used.

The following instructions show the keys for entering the main program. To make a correction, move the cursor to the error and use $\boxed{\text{INS}}$ and $\boxed{\text{DEL}}$ as needed. To return to the program from other screens, press $\boxed{\text{QUIT}}$ or follow the steps in the previous paragraph. If that does not work or you want to stop the running of the program, press $\boxed{\text{ON}}$ $\boxed{2}$.

---

1. $\boxed{\text{R}}$ $\boxed{\text{U}}$ $\boxed{\text{N}}$ $\boxed{\text{G}}$ $\boxed{\text{E}}$ $\boxed{\text{ENTER}}$. This title identifies the program.

---

2. $\boxed{\text{2nd}}$ $\boxed{\text{Y-VARS}}$ $\boxed{\blacktriangleleft}$ $\boxed{1}$ $\boxed{\text{ENTER}}$. All-off unselects all the formulas in the Y= screen so their graphs will not be generated when the program is run.

---

3. $\boxed{\text{PRGM}}$ $\boxed{\blacktriangleright}$ $\boxed{1}$ $\boxed{\text{2nd}}$ $\boxed{\text{A-LOCK}}$ $\boxed{\text{``}}$ $\boxed{\text{U}}$ $\boxed{\text{S}}$ $\boxed{\text{E}}$ $\boxed{\text{``}}$ $\boxed{\text{ENTER}}$.

---

4. $\boxed{\text{PRGM}}$ $\boxed{\blacktriangleright}$ $\boxed{1}$ $\boxed{\text{2nd}}$ $\boxed{\text{A-LOCK}}$ $\boxed{\text{``}}$ $\boxed{D}$ $\boxed{V}$ $\boxed{\text{ALPHA}}$ $\boxed{\div}$ $\boxed{\text{ALPHA}}$ $\boxed{D}$ $\boxed{X|T}$ $\boxed{\text{2nd}}$ $\boxed{\text{TEST}}$ $\boxed{1}$ $\boxed{\text{2nd}}$ $\boxed{\text{Y-vars}}$ $\boxed{1}$ $\boxed{(}$ $\boxed{\text{2nd}}$ $\boxed{\text{A-LOCK}}$ $\boxed{X|T}$ $\boxed{,}$ $\boxed{V}$ $\boxed{\text{ALPHA}}$ $\boxed{)}$ $\boxed{\text{ALPHA}}$ $\boxed{\text{``}}$ $\boxed{\text{ENTER}}$.

---

5. $\boxed{\text{PRGM}}$ $\boxed{\blacktriangleright}$ $\boxed{1}$ $\boxed{\text{2nd}}$ $\boxed{\text{A-LOCK}}$ $\boxed{\text{``}}$ $\boxed{A}$ $\boxed{\text{2nd}}$ $\boxed{\text{TEST}}$ $\boxed{1}$ $\boxed{\text{ALPHA}}$ $\boxed{\text{``}}$ $\boxed{\text{ENTER}}$.

---

6. [PRGM] [▶] [2] [ALPHA] [A] [ENTER] .

7. [PRGM] [▶] [1] [2nd] [A-LOCK] ["] [B] [2nd] [TEST] [1] [ALPHA] ["] [ENTER] .

8. [PRGM] [▶] [2] [ALPHA] [B] [ENTER] .

9. [PRGM] [▶] [1] [2nd] [A-LOCK] ["] [V] [ALPHA] [∅] [2nd] [TEST] [1] [ALPHA] ["] [ENTER] .

10. [PRGM] [▶] [2] [ALPHA] [V] [ENTER] .

11. [PRGM] [▶] [1] [2nd] [A-LOCK] ["] [N] [2nd] [TEST] [1] [ALPHA] ["] [ENTER] .

12. [PRGM] [▶] [2] [ALPHA] [N] [ENTER] .

13. [(] [ALPHA] [B] [−] [ALPHA] [A] [)] [÷] [ALPHA] [N] [STO▶] [D] [ENTER] .

14. [ALPHA] [A] [STO▶] [T] [ENTER] .

15. [PRGM] [1] [1] [ENTER] .

16. [ALPHA] [T] [STO▶] [C] [ENTER] .

17. [ALPHA] [V] [STO▶] [Z] [ENTER] .

18. [ALPHA] [C] [STO▶] [X|T] [ENTER] .

19. [2nd] [Y-VARS] [1] [STO▶] [P] [ENTER] .

20. [ALPHA] [C] [+] [ALPHA] [D] [÷] [2] [STO▶] [X|T] [ENTER] .

Graphing calculator workbook

21. [ALPHA] [Z] [+] [ALPHA] [D] [ALPHA] [P] [÷] [2] [STO▶] [V] [ENTER] .

22. [2nd] [Y-VARS] [1] [STO▶] [Q] [ENTER] .

23. [ALPHA] [Z] [+] [ALPHA] [D] [ALPHA] [Q] [÷] [2] [STO▶] [V] [ENTER] .

24. [2nd] [Y-VARS] [1] [STO▶] [R] [ENTER] .

25. [ALPHA] [C] [+] [ALPHA] [D] [STO▶] [X|T] [ENTER] .

26. [ALPHA] [Z] [+] [ALPHA] [D] [ALPHA] [R] [STO▶] [V] [ENTER] .

27. [2nd] [Y-VARS] [1] [STO▶] [S] [ENTER] .

28. [(] [ALPHA] [P] [+] [2] [ALPHA] [Q] [+] [2] [ALPHA] [R] [+] [ALPHA] [S] [)] [÷] [6] [STO▶] [M] [ENTER] .

29. [X|T] [STO▶] [T] [ENTER] .

30. [ALPHA] [Z] [+] [ALPHA] [D] [ALPHA] [M] [STO▶] [V] [ENTER] .

31. [2nd] [DRAW] [2] [2nd] [A-LOCK] [C] [,] [Z] [,] [T] [,] [V] [ALPHA] [)] [ENTER] .

32. PRGM | 3 | ALPHA | T | 2nd | TEST | 5 | ALPHA | B | ENTER .

33. PRGM | 2 | 1 | ENTER .

34. PRGM | 6 | ENTER .

35. PRGM | ◄ | ALPHA | P | ENTER .

36. PRGM | 3 | ALPHA | θ | 2nd | TEST | 2 | 0 | ENTER .

37. PRGM | 6 | ENTER .

38. PRGM | ► | 1 | 2nd | A-LOCK | " | T | H | E | - | M | E | T | H | O | D | - | G | I | V | E | S | " | ENTER .

39. PRGM | ► | 1 | 2nd | A-LOCK | " | V | 2nd | TEST | 1 | ALPHA | " | ENTER .

40. PRGM | ► | 1 | ALPHA | V | ENTER .

41. PRGM | ► | 1 | 2nd | A-LOCK | " | A | T | - | X|T | 2nd | TEST | 1 | ALPHA | " | ENTER .

42. PRGM | ► | 1 | ALPHA | B | ENTER .

Using the program

Example   Use the Runge-Kutta program 5D with $N = 10$ to generate the graph of the approximate solution of the initial value problem

$$\frac{dv}{dx} = \tfrac{1}{4} v^2 \sin x, \quad v(0) = \tfrac{4}{3} \tag{2}$$

for $0 \leq x \leq 10$. Use $-1 \leq x \leq 11, -0.5 \leq y \leq 5$ for the graph. Then generate on the same screen the graph of the exact solution $4/(2 + \cos x)$ of (2) and compare the values of the approximate and exact solutions at $x = 10$.

Solution   Enter the function $\tfrac{1}{4} v^2 \sin x$ from the differential equation as $Y_1 = (1/4)V^2 \sin X$ and the exact solution as $Y_2 = 4/(2 + \cos X)$ in the Y= menu and set x-min $= -1$, x-max $= 11$, x-scale $= 1$, y-min $= -0.5$, y-max $= 5$, and y-scale $= 1$ in the range

Graphing calculator workbook

menu. The graphics screen is cleared automatically if you change the Y= or range menus. Otherwise clear it.

Then press [PRGM] [6] (or whatever number you gave this program) [ENTER]. The symbols "USE DV/DX = $Y_1(X,V)$" will appear to remind you that x and v should be the variables and the function on the right side of the differential equation should be entered as $Y_1(X,V)$. After the prompt "$A =?$", press [0] [ENTER] to set $A = 0$. After the next prompts use [1] [0] [ENTER] to set $B = 10$, [4] [÷] [3] [ENTER] to set $v_0 = \frac{4}{3}$, and [1] [0] [ENTER] to set $N = 10$.

Press [ENTER] to see the graph of the approximate solution as in Figure 6.
Press [ENTER] and then [2] [ENTER] after the prompt "0 FOR NO PLOT, K TO PLOT YK" to add the graph of the exact solution $v(x) = 4/(2+\cos x)$ as in Figure 7. Press [ENTER] [0] [ENTER] to display the value of the approximate solution at $B = 10$ and exit the program.

The Runge Kutta method gives the approximate value 3.34754432 for $v(10)$. Press [1] [0] [STO▶] [X|T] to store 10 as X, and [2nd] [Y-VARS] [2] [ENTER] for the value 3.445518049 of the exact solution at $x = 10$. It differs from the value of the Runge-Kutta approximation by approximately 0.098. Greater accuracy could be achieved with a larger N.

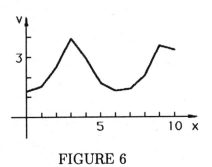

FIGURE 6

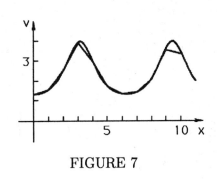

FIGURE 7

The above program can be modified to carry out Euler's method as well as the Runge Kutta method by inserting

4a. : Disp " GIVE 1 FOR EULER"
4b. : Disp " 2 FOR RUNGEKUTTA"
4c. : Input L

after line 4 and

28a. : If $L = 1$
28b. : $P \to M$

after line 28.

Graphing calculator workbook 217

Worksheet 6A.1[†]

Going part way

Name _____ Date _____

Others in your group _____

Instructor, Teaching Assistant, and/or Recitation section _____

Directions *Put first drafts of your calculations and answers on scratch paper. Take your time, work carefully, and discuss your solution with at least one other student before putting a final draft on this sheet or on other paper. Turn in all your work.*

Problem 1 Suppose you want to go from one point A toward a second point B that is two miles away. First you go one mile; then you go $\frac{1}{2}$ mile farther; next you go half that distance, which is an additional $\left(\frac{1}{2}\right)^2 = \frac{1}{4}$ mile; then half that distance, which is an additional $\left(\frac{1}{2}\right)^3 = \frac{1}{8}$ mile, and so forth.

  a. As is indicated in Table 1, you go 1 mile in one step and 1.5 mile in two steps. Complete the table by calculating how far you go in three, five, and ten steps.

  b. How far do you think you would go in an infinite number of steps? If you cannot guess the answer right away, calculate how far you would go in 15 or 20 steps.

Steps	Distance	Decimal value
1	1	1.0
1–2	$1 + \frac{1}{2}$	1.5
1–3	$1 + \frac{1}{2} + \left(\frac{1}{2}\right)^2$	
1–5	$1 + \frac{1}{2} + \left(\frac{1}{2}\right)^2 + \left(\frac{1}{2}\right)^3 + \left(\frac{1}{2}\right)^4$	
1–10	$1 + \frac{1}{2} + \left(\frac{1}{2}\right)^2 + \left(\frac{1}{2}\right)^3 + \cdots + \left(\frac{1}{2}\right)^9$	

TABLE 1

Steps	Distance	Decimal value
1	1	1.0
1–2	$1 + \frac{1}{4}$	1.25
1–3	$1 + \frac{1}{4} + \left(\frac{1}{4}\right)^2$	
1–5	$1 + \frac{1}{4} + \left(\frac{1}{4}\right)^2 + \left(\frac{1}{4}\right)^3 + \left(\frac{1}{4}\right)^4$	
1–10	$1 + \frac{1}{4} + \left(\frac{1}{4}\right)^2 + \left(\frac{1}{4}\right)^3 + \cdots + \left(\frac{1}{4}\right)^9$	

TABLE 2

[†]This worksheet is an introduction to geometric series.

218   Graphing calculator workbook

Problem 2   Now suppose you go one mile in the first step, one-fourth mile in the second step, and in each subsequent step one-fourth as far as in the previous step.

   a.   Complete Table 2 to show how far you go in three, five, and ten steps.

   b.   How far do you think you would go in an infinite number of steps? If you cannot guess the answer right away, calculate how far you would go in 15 or 20 steps.

Problem 3   This time suppose you go one mile from A toward B in the first step, but then $\frac{1}{2}$ mile backwards toward A in the second step, $\left(\frac{1}{2}\right)^2 = \frac{1}{4}$ mile forward in the third step, $\left(\frac{1}{2}\right)^3 = \frac{1}{8}$ mile backward in the fourth step, and so on.

   a.   Complete Table 3 to show how far you go in three, five, and ten steps.

   b.   How far do you think you would go in an infinite number of steps?

Steps	Distance	Decimal value
1	1	1.0
1–2	$1 - \frac{1}{2}$	0.5
1–3	$1 - \frac{1}{2} + \left(\frac{1}{2}\right)^2$	
1–5	$1 - \frac{1}{2} + \left(\frac{1}{2}\right)^2 - \left(\frac{1}{2}\right)^3 + \left(\frac{1}{2}\right)^4$	
1–10	$1 - \frac{1}{2} + \left(\frac{1}{2}\right)^2 - \left(\frac{1}{2}\right)^3 + \cdots - \left(\frac{1}{2}\right)^9$	

TABLE 3

Problem 4   Problems 1 through 3 can be solved more quickly by using the finite geometric series formula

$$1 + r + r^2 + r^3 + \cdots + r^{n-1} = \frac{1 - r^n}{1 - r} \tag{1}$$

which is valid for any positive integer n and any number $r \neq 1$. (Check it for, say $n = 5$, by expanding and simplifying the expression $(1 - r)(1 + r + r^2 + r^3 + r^4)$ that is obtained by multiplying the left sides of (1) by $(1 - r)$.)

   a.   Use (1) with $r = \frac{1}{2}$ to check your numbers in Table 1, with $r = \frac{1}{4}$ to check Table 2, and with $r = -\frac{1}{2}$ to check Table 3.

   b.   Use (1) to answer parts (b) of Problems 1 through 3. (Use the fact that $\left(\frac{1}{2}\right)^n$ and $\left(\frac{1}{4}\right)^n$ tend to zero as n tends to ∞.)

Graphing calculator workbook   219

Worksheet 6A.2[†]

Loan payments and St. Ives

Name _____ Date _____

Others in your group _____

Instructor, Teaching Assistant, and/or Recitation section _____

Directions   *Put first drafts of your calculations and answers on scratch paper. Take your time, work carefully, and discuss your solution with at least one other student before putting a final draft on this sheet or on other paper. Turn in all your work.*

Problem 1   Suppose you borrow $1,000 at an interest rate of 12% per year (1% per month), and you make payments of $200 per month, starting one month after you take out the loan. The $1,000.00 in column 4 of the line for month 0 in the table below is the amount of the loan. The second line shows a calculation of the status of your loan after the first payment. The $1,010.00 in the second column consists of the previous balance of $1,000 plus one-month's interest of 1% of that balance. Your $200 payment then reduces the balance to $810, as shown in the last column. The loan is paid off with the sixth payment.

    a.   Carry out the calculations to fill in the missing numbers.

    b.   How much total interest do you pay for the loan?

    c.   How much interest would you have had to pay at the same interest rate if you had repaid the loan with a lump sum at the end of six months instead of with monthly payments? (The answer is larger than the answer to part (b).)

Month	Balance with interest before the payment	Payment	Balance after the payment
0			$1,000.00
1	$1,010.00	$200	$810.00
2		$200	
3		$200	
4		$200	
5		$200	
6	$31.12	$31.12	$0.00

_____

    [†]This worksheet shows how geometric series can be used to derive formulas for loan balances, and how a nursery rhyme cntains a geometric series.

Graphing calculator workbook

Problem 2   The results of the repetitive calculations used to complete Table 1 can be represented by concise formulas which can then be applied more quickly with less risk of error. We let B_n denote the loan balance after the nth payment—the numbers in the last column of the table—so that $B_0 = \$1000.00$. We let P denote the monthly payment of $200.

a.   Explain why $B_1 = (1.01)B_0 - P, B_2 = (1.01)B_1 - P, B_3 = (1.01)B_2 - P$, $B_4 = (1.01)B_3 - P$, and $B_5 = (1.01)B_4 - P$.

b.   Combine the formulas from part (a) to show that

$$B_2 = (1.01)^2 B_0 - (1.01)P - P.$$

c.   Express B_3, B_4 and B_5 in terms of B_0 and P. The last result should be

$$B_5 = (1.01)^5 B_0 - \left[1 + 1.01 + (1.01)^2 + (1.01)^3 + (1.01)^4\right]P. \tag{1}$$

d.   Eliminate the sum in front of P in (1) by using the finite geometric series formula,

$$1 + r + r^2 + r^3 + r^4 = \frac{1 - r^5}{1 - r} \tag{2}$$

which is valid for $r \neq 1$.

e.   Use the formula from part (d) to check the value of B_5 in the last column of Table 1.

Problem 3   An eighteenth century nursery rhyme (slightly modified), says, "As I was going to St. Ives, I met a man with seven wives. Every wife had seven sacks, every sack had seven cats, every cat had seven kits, every kit had seven mice, and every mouse had seven lice. Lice, mice, kits, cats, sacks, wives—how many were going to St. Ives?"

a.   Answer the question.

b.   How many were coming from St. Ives? Use the finite geometric series formula

$$1 + r + r^2 + r^3 + \cdots + r^n = \frac{1 - r^{n+1}}{1 - r} \tag{3}$$

which is valid for $r \neq 1$ and any positive integer n.

Graphing calculator workbook   221

Worksheet 6A.3[†]

A flip of a coin, a roll of the dice

Name _____ Date _____

Others in your group _____

Instructor, Teaching Assistant, and/or Recitation section _____

Directions   *Put first drafts of your calculations and answers on scratch paper. Take your time, work carefully, and discuss your solution with at least one other student before putting a final draft on this sheet or on other paper. Turn in all your work.*

Problem 1   The set of possible outcomes from an experiment is called the *sample space* of the experiment. For example, the sample space of flipping a coin is the set

$$\boxed{\text{H}} \qquad \boxed{\text{T}}$$

where $\boxed{\text{H}}$ denotes heads and $\boxed{\text{T}}$ denotes tails. We assume that the coin is fair. Then the probability of getting a head is $\frac{1}{2}$ and the probability of getting a tail is $\frac{1}{2}$. The sample space of flipping the coin twice can be represented by

$$\boxed{\text{H H}} \qquad \boxed{\text{H T}}$$
$$\boxed{\text{T H}} \qquad \boxed{\text{T T}}$$

where the first letter in each box is the result of the first flip and the second letter is the result of the second flip. We list $\boxed{\text{H T}}$ and $\boxed{\text{T H}}$ separately so that each outcome in the list has probability $\frac{1}{4}$ of occurring.

a.   What is the probability of getting two heads in two flips of the coin? Of getting one head and one tail (in either order)? Of getting two tails? What is the sum of these three probabilities?

b.   The sample space of flipping the coin three times is

$$\boxed{\text{H H H}} \qquad \boxed{\text{H H T}} \qquad \boxed{\text{H T H}} \qquad \boxed{\text{H T T}}$$
$$\boxed{\text{T H H}} \qquad \boxed{\text{T H T}} \qquad \boxed{\text{T T H}} \qquad \boxed{\text{T T T}}$$

where the first letter is the result of the first flip, the second is the result of the second flip, and the third is the result of the third flip. Here $\boxed{\text{H H T}}$ is listed separately from $\boxed{\text{H T H}}$ and $\boxed{\text{T H H}}$ so that each of the eight possible outcomes is equally likely. What is the probability of getting three heads? Of getting two heads and one tail, in any order? Of getting one head and two tails, in any order? Of getting three tails? What is the sum of these probabilities?

[†]This worksheet discusses an application of geometric series to probability.

222   Graphing calculator workbook

Problem 2a Imagine that the coin is flipped an infinite number of times. What are the probabilities that the first flip is a head; that the first flip is a tail and the second is a head; that the first two flips are tails and the third is a head; and that the first three flips are tails and the fourth is a head?

b. What is the probability that the first head occurs in the jth flip, where j is an arbitrary positive integer?

c. Explain why the probability that a head occurs first in one of the first k flips is given for any positive integer k by the sum

$$\tfrac{1}{2} + \left(\tfrac{1}{2}\right)^2 + \left(\tfrac{1}{2}\right)^3 + \cdots + \left(\tfrac{1}{2}\right)^k = \tfrac{1}{2}\left[1 + \tfrac{1}{2} + \left(\tfrac{1}{2}\right)^2 + \left(\tfrac{1}{2}\right)^3 + \cdots + \left(\tfrac{1}{2}\right)^{k-1}\right]. \qquad (1)$$

d. Use the finite geometric series formula

$$1 + r + r^2 + r^3 + \cdots + r^{k-1} = \frac{1 - r^k}{1 - r} \qquad (2)$$

to find the limit of (1) as $k \to \infty$. (Factor out $\tfrac{1}{2}$). What does this tell you about the probabilities?

Problem 3 The sample space of rolling a die once is

$$\boxed{1} \quad \boxed{2} \quad \boxed{3} \quad \boxed{4} \quad \boxed{5} \quad \boxed{6}$$

where the number denotes the number of dots on the top face of the die. If the die is fair, each of these six events is equally likely and each events probability $\tfrac{1}{6}$ of occuring.

   The sample space for rolling the die twice is a set of thirty-six equally likely events, including the following six:

$$\boxed{1\ \ 1} \quad \boxed{1\ \ 2} \quad \boxed{1\ \ 3} \quad \boxed{1\ \ 4} \quad \boxed{1\ \ 5} \quad \boxed{1\ \ 6}$$

a. List the other 30 possibilities.

b. Suppose the die is rolled twice and the number of dots on the top faces are added, yielding a total between 2 and 12. What is the probability that each of these eleven totals occurs? What is the sum of these probabilities?

Problem 4 Suppose that the die is rolled an infinite number of times.

a. What is the probability that a 1 occurs in the first roll? What is the probability that a 1 does not occur in the first roll but does occur in the second? Explain why the probability that a 1 does not occur in the first or second roll but does occur in the third is $\left(\tfrac{1}{6}\right)\left(\tfrac{5}{6}\right)^2$.

b. Explain why the probability that a 1 occurs for the first time in the jth roll is $\left(\tfrac{1}{6}\right)\left(\tfrac{5}{6}\right)^{j-1}$.

c. Explain why the probability that a head occurs first in one of the first k rolls is

$$\tfrac{1}{6} + \tfrac{1}{6}\left(\tfrac{5}{6}\right) + \tfrac{1}{6}\left(\tfrac{5}{6}\right)^2 + \tfrac{1}{6}\left(\tfrac{5}{6}\right)^3 + \cdots + \tfrac{1}{6}\left(\tfrac{5}{6}\right)^{k-1} \qquad (3)$$

d. Use the finite geometric series formula (2) to find the limit of (3) as $k \to \infty$. Explain the result.

Graphing calculator workbook   223

Sequence program 6B

Texas Instruments TI-81

This program draws a bar graph showing the values of a sequence $\{a_J\}$ from $J = J_0$ to Max$J$, where a_J is a function of J given as Y_1 in the Y= menu and suitable ranges of x and y have been chosen for the graph. The user supplies J_0 and Max$J$. Subprogram P, that is listed with the direction field program 5B, should also be entered. It allows the user to generate graphs of functions of x given as Y_2, Y_3 or Y_4 in the Y= menu after the sequence is plotted.

The main program (103 bytes)

1.   Prgm7: SEQUENCE
2.   : ClrDraw
3.   : All-Off
4.   : Disp "USE $AJ = Y_1(J)$"
5.   : Disp "$J\emptyset =$"   ($\emptyset$ denotes zero.)
6.   : Input A
7.   : Disp "MAX J $=$"
8.   : Input N
9.   : $A \rightarrow J$
10.   : Lbl 1
11.   : Disp "$J =$"
12.   : Disp J
13.   : $Y_1 \rightarrow Z$
14.   : Disp "$AJ =$"
15.   : Disp Z
16.   : Pause
17.   : Line($J, \emptyset, J, Z$)
18.   : Pause
19.   : $J + 1 \rightarrow J$
20.   : If $J \leq N$
21.   : Goto 1
22.   : PrgmP

Entering the main program

---

1. Press $\boxed{S}$ $\boxed{E}$ $\boxed{Q}$ $\boxed{U}$ $\boxed{E}$ $\boxed{N}$ $\boxed{C}$ $\boxed{E}$.

---

2. Press $\boxed{\text{2nd}}$ $\boxed{\text{DRAW}}$ $\boxed{1}$ $\boxed{\text{ENTER}}$. ClrDraw clears the graphics screen.

---

3. Press $\boxed{\text{2nd}}$ $\boxed{\text{Y-VARS}}$ $\boxed{\blacktriangleleft}$ $\boxed{1}$ $\boxed{\text{ENTER}}$. All-Off unselects the functions in the Y= menu so their graphs will not be generated.

---

4. Press $\boxed{\text{PRGM}}$ $\boxed{\blacktriangleright}$ $\boxed{1}$ $\boxed{\text{2nd}}$ $\boxed{\text{A-LOCK}}$ $\boxed{``}$ $\boxed{\text{U}}$ $\boxed{\text{S}}$ $\boxed{\text{E}}$ $\boxed{\text{-}}$ $\boxed{\text{A}}$ $\boxed{\text{J}}$ $\boxed{\text{2nd}}$ $\boxed{\text{TEST}}$ $\boxed{1}$ $\boxed{\text{2nd}}$ $\boxed{\text{Y-VARS}}$ $\boxed{1}$ $\boxed{(}$ $\boxed{\text{ALPHA}}$ $\boxed{\text{J}}$ $\boxed{)}$ $\boxed{\text{ALPHA}}$ $\boxed{``}$ $\boxed{\text{ENTER}}$.

5. Press $\boxed{\text{PRGM}}$ $\boxed{\blacktriangleright}$ $\boxed{1}$ $\boxed{\text{ALPHA}}$ $\boxed{``}$ $\boxed{\text{ALPHA}}$ $\boxed{\text{J}}$ $\boxed{\emptyset}$ $\boxed{\text{2nd}}$ $\boxed{\text{TEST}}$ $\boxed{1}$ $\boxed{\text{ALPHA}}$ $\boxed{``}$ $\boxed{\text{ENTER}}$.

6. Press $\boxed{\text{PRGM}}$ $\boxed{\blacktriangleright}$ $\boxed{2}$ $\boxed{\text{ALPHA}}$ $\boxed{A}$ $\boxed{\text{ENTER}}$. Here the user assigns a value to $J0$, which is stored as A.

7. Press $\boxed{\text{PRGM}}$ $\boxed{\blacktriangleright}$ $\boxed{1}$ $\boxed{\text{2nd}}$ $\boxed{\text{A-LOCK}}$ $\boxed{``}$ $\boxed{\text{M}}$ $\boxed{\text{A}}$ $\boxed{\text{X}}$ $\boxed{\text{-}}$ $\boxed{\text{J}}$ $\boxed{\text{2nd}}$ $\boxed{\text{TEST}}$ $\boxed{1}$ $\boxed{\text{ALPHA}}$ $\boxed{``}$ $\boxed{\text{ENTER}}$.

8. Press $\boxed{\text{PRGM}}$ $\boxed{\blacktriangleright}$ $\boxed{2}$ $\boxed{\text{ALPHA}}$ $\boxed{N}$ $\boxed{\text{ENTER}}$. Set $N = MAXJ$.

9. Press $\boxed{\text{ALPHA}}$ $\boxed{A}$ $\boxed{\text{STO}\blacktriangleright}$ $\boxed{J}$ $\boxed{\text{ENTER}}$ to set J equal to $A = J0$.

10. Press $\boxed{\text{PRGM}}$ $\boxed{1}$ $\boxed{1}$ $\boxed{\text{ENTER}}$. The operation of the program will return to this label when it reaches the Goto 1 that will be put on line 21.

11. Press $\boxed{\text{PRGM}}$ $\boxed{\blacktriangleright}$ $\boxed{1}$ $\boxed{\text{ALPHA}}$ $\boxed{``}$ $\boxed{\text{ALPHA}}$ $\boxed{\text{J}}$ $\boxed{\text{2nd}}$ $\boxed{\text{TEST}}$ $\boxed{1}$ $\boxed{\text{ALPHA}}$ $\boxed{``}$ $\boxed{\text{ENTER}}$. Display $J =$.

12. Press $\boxed{\text{PRGM}}$ $\boxed{\blacktriangleright}$ $\boxed{1}$ $\boxed{\text{ALPHA}}$ $\boxed{\text{J}}$ $\boxed{\text{ENTER}}$. Display the value of J.

13. Press $\boxed{\text{2nd}}$ $\boxed{\text{Y-VARS}}$ $\boxed{1}$ $\boxed{\text{STO}\blacktriangleright}$ $\boxed{Z}$ $\boxed{\text{ENTER}}$. Set $Z = Y_1(J)$.

14. Press $\boxed{\text{PRGM}}$ $\boxed{\blacktriangleright}$ $\boxed{1}$ $\boxed{\text{2nd}}$ $\boxed{\text{A-LOCK}}$ $\boxed{``}$ $\boxed{\text{A}}$ $\boxed{\text{J}}$ $\boxed{\text{2nd}}$ $\boxed{\text{TEST}}$ $\boxed{1}$ $\boxed{\text{ALPHA}}$ $\boxed{``}$ $\boxed{\text{ENTER}}$ to have the program display $AJ =$.

15. Press $\boxed{\text{PRGM}}$ $\boxed{\blacktriangleright}$ $\boxed{1}$ $\boxed{\text{ALPHA}}$ $\boxed{Z}$ $\boxed{\text{ENTER}}$. Display a_J's value.

16. Press $\boxed{\text{PRGM}}$ $\boxed{6}$ $\boxed{\text{ENTER}}$. The program pauses to show the values.

17. Press $\boxed{\text{2nd}}$ $\boxed{\text{DRAW}}$ $\boxed{2}$ $\boxed{\text{ALPHA}}$ $\boxed{\text{J}}$ $\boxed{\text{ALPHA}}$ $\boxed{,}$ $\boxed{\emptyset}$ $\boxed{\text{2nd}}$ $\boxed{\text{A-LOCK}}$ $\boxed{,}$ $\boxed{\text{J}}$ $\boxed{,}$ $\boxed{\text{Z}}$ $\boxed{\text{ALPHA}}$ $\boxed{)}$ $\boxed{\text{ENTER}}$. Draw a vertical line from $(J, 0)$ on the x-axis to $(J, Z) = (J, a_J)$.

Graphing calculator workbook   225

18. Press [PRGM] [6] [ENTER]. The program pauses to show the graph.

19. Press [ALPHA] [J] [+] [1] [STO▶] [J] [ENTER] to increase the value of J by 1.

20. Press [PRGM] [3] [ALPHA] [J] [2nd] [TEST] [6] [ALPHA] [N] [ENTER]. If J is $\leq N =$ Max$J$, the next line is executed and the loop is repeated. Otherwise the next line is skipped and the program ends.

21. Press [PRGM] [2] [1] [ENTER].

22. Press [PRGM] [◀] [ALPHA] [P] [ENTER]. This command begins the execution of subprogram P.

Using the program

Example a. Use the sequence program to create a bar graph showing the first fifteen values of the sequence

$$\{a_j\}_{j=1}^{\infty} = \left\{\frac{j}{50+j^3}\right\}_{j=1}^{\infty}. \tag{1}$$

b. What is the limit of a_j as $j \to \infty$? Use algebra to justify your answer.

c. What is the maximum value of the sequence $\{a_j\}$? Use calculus to justify your answer.

Solution a. Press [Y=] and enter the formula for the sequence as $Y_1 = J/(50+J \wedge 3)$. Set Xmin $= -1$ and Xmax $= 16$ to show the graph for $1 \leq J \leq 15$. After some experimentation, we decide to use $-0.01 \leq y \leq 0.05$ with y-scale $= 0.01$.

Press [PRGM], the number of your sequence method program, and [ENTER]. When the prompt $J0 =?$ appears, press [1] [ENTER] and when MAX $J =?$ appears, press [15] [ENTER] to give J_0 the value 1 and Max$J$ the value 15.

We see that the first number in the sequence a_1 has the approximate numerical value 0.0196078431. Continuing to press [ENTER] gives the graph in Figure 1 and the values

$a_1 \doteq 0.0196078431,$ $a_2 \doteq 0.0344827586,$ $a_3 \doteq 0.038961039$
$a_4 \doteq 0.0350877193,$ $a_5 \doteq 0.0285714286,$ $a_6 \doteq 0.0225563910$
$a_7 \doteq 0.0178117048,$ $a_8 \doteq 0.0142348754,$ $a_9 \doteq 0.0115532734$
$a_{10} \doteq 0.0095238095,$ $a_{11} \doteq 0.0079652426,$ $a_{12} \doteq 0.0067491564$
$a_{13} \doteq 0.0057854918,$ $a_{14} \doteq 0.0050107373,$ $a_{15} \doteq 0.004379562$

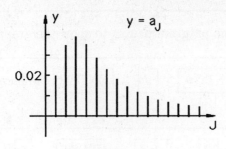
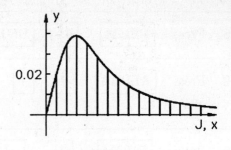

FIGURE 1   FIGURE 2

Press $\boxed{0}$ $\boxed{\text{ENTER}}$ after the prompt "0 for no plot, K to plot YK" or press $\boxed{\text{ENTER}}$ $\boxed{\text{ON}}$ $\boxed{2}$ to stop the program.

b. The table of values and the graph suggest that $a_j \to \infty$ as $j \to \infty$. It is not obvious, however, from the given formula what actually happens to a_j as $j \to \infty$ because the numerator and denominator both tend to ∞. Because the numerator has lower degree, we divide by its highest power of j, which is j itself, and obtain

$$a_j = \frac{j}{50 + j^3} = \frac{1}{\frac{50}{j} + j^2} \quad \text{for } j \neq 0$$

Since the new denominator $\frac{50}{j} + j^2$ tends to ∞ as $j \to \infty$ and the new numerator is constant, a_j does tends to 0 as $j \to \infty$.

c. The table and graph suggest that the greatest value of the sequence is $a_3 = \frac{3}{50 + 3^3} = \frac{3}{77} \doteq 0.0389610389$, but we do not yet know what anything about a_j for $j > 16$. We use calculus to find the maximum value of the function $f(x) = \frac{x}{50 + x^3}$, defined for all positive values of x. Its derivative is

$$\frac{d}{dx} \frac{x}{50 + x^3} = \frac{(50 + x^3)\frac{d}{dx}(x) - x\frac{d}{dx}(50 + x^3)}{(50 + x^3)^2}$$

$$= \frac{50 + x^3 - 3x^3}{(50 + x^3)^2} = \frac{2(25 - x^3)}{(50 - x^3)^2}.$$

and is zero at $x = 25^{1/3}$. Figure 3 shows the graph of $f(x)$ with the bar graph of Figure 2. Because the derivative of $f(x)$ is positive for $0 < x < 25^{1/3}$ and negative for $x > 25^{1/3}$, $f(x)$ is increasing for $0 < x < 25^{1/3}$ and decreasing for $x > 25^{1/3}$ and its maximum value for $x > 0$ is its value at $25^{1/3} \doteq 2.924$. Because the integers to the left and to the right of $25^{1/3}$ are 2 and 3, the maximum a_j is either a_2 or a_3. The graph or table show that it is a_3, as we expected.

To obtain Figure 2, set $Y_2 = X/(50 + X \wedge 3)$, run the program again, and give 2 after the last prompt.

Worksheet 6C.1†

Infinite sequences: Initial investigations

Name _____ Date _____

Others in your group _____

Instructor, Teaching Assistant, and/or Recitation section _____

Directions   Put first drafts of your calculations and answers on scratch paper. Take your time, work carefully, and discuss your solution with at least one other student before putting a final draft on this sheet or on other paper. Turn in all your work.

Problem 1   Use Sequence program 6B with $-1 \le x \le 25$ and x-scale $= 1$, but without picking a y-range, to calculate

$$a_j = \frac{6}{1+j}$$

for $j = 1, 2, 3, \ldots 24$. Then use the displayed values to pick a suitable y-range and run the program again. Copy the resulting bar graph in Figure 1 and label the scale on the y-axis. Describe the behavior of the sequence. For instance, does it increase or decrease or oscillate for any ranges of j? Does it have a finite or infinite limit as $j \to \infty$? Does it have a least or greatest value? Does it take any values infinitely often? What other distinctive characteristics does it have?

Problem 2   Follow the instructions of Problem 1 for $a_j = 1.5 + (-1)^{j+1}$. Use Figure 2.

Problem 3   Follow the instructions of Problem 1 with $a_j = j^{1/3}$. Use Figure 3.

Problem 4   Follow the instructions of Problem 1 for $a_j = \dfrac{16j - j^2}{14 + j}$. Use Figure 4.

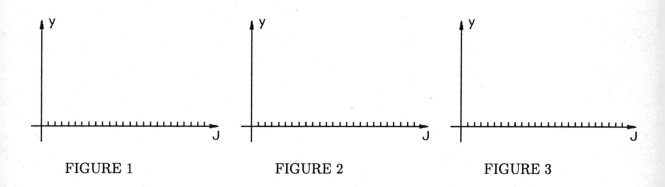

FIGURE 1   FIGURE 2   FIGURE 3

---

†This worksheet uses Sequence program 3B to illustrate some of the different types of behavior of infinite sequences.

Problem 5 Follow the instructions of Problem 1 for $a_j = j + 2(-1)^j$. Use Figure 5.

Problem 6 Follow the instructions of Problem 1 for $a_j = \cos\left(\dfrac{j\pi}{3}\right)$. Use Figure 6.

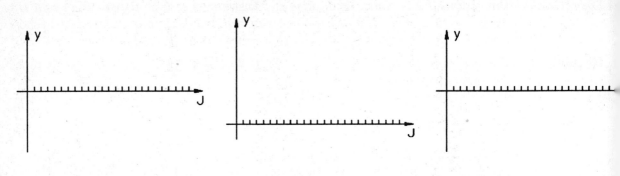

FIGURE 4　　　　　　　　　　FIGURE 5　　　　　　　　　　FIGURE 6

Problem 7 Follow the instructions of Problem 1 with $a_j = \dfrac{j}{j - 8.5}$. Use Figure 7.

Problem 8 Follow the instructions of Problem 1 for $a_j = \left(1 + \dfrac{1}{j}\right)^j$. Use Figure 8.

Problem 9 Follow the instructions of Problem 1 with $a_j = \dfrac{\ln j}{j}$. Use Figure 9.

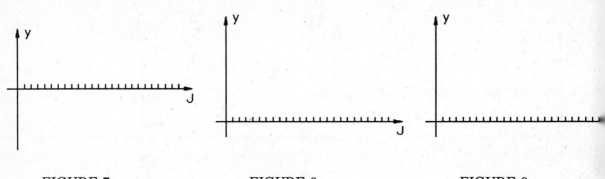

FIGURE 7　　　　　　　　　　FIGURE 8　　　　　　　　　　FIGURE 9

Worksheet 6C.2†

The ϵN- and YN-definitions of convergence and divergence to ∞

Name _____ Date _____

Others in your group _____

Instructor, Teaching Assistant, and/or Recitation section _____

Directions Put first drafts of your calculations and answers on scratch paper. Take your time, work carefully, and discuss your solution with at least one other student before putting a final draft on this sheet or on other paper. Turn in all your work.

Problem 1 The sequence $a_j = 2 + \dfrac{\cos(j\pi/3)}{j^2}$ $(j = 1, 2, 3, \dots)$ tends to 2 as j tends to ∞ because

$$|\cos(j\pi/3)| \le 1 \text{ for all } j. \qquad (1)$$

and $1/j^2 \to 0$. According to the ϵN-definition of this limit, there is, for every positive ϵ, no matter how small, a positive N_ϵ such that $|a_j - 2| < \epsilon$ for $j > N_\epsilon$.

a. Use (1) to find an N_ϵ for $\epsilon = 0.1$. As a partial check of your result, use Sequence program 6B to plot a_j for $j = 1, 2, 3, \dots, 12$ as a bar graph with the lines $y = 2.1$ and $y = 1.9$ for $-1 \le x \le 13, -0.5 \le y \le 3$. Copy the graph and lines in Figure 1 and explain how they relate to your choice of N_ϵ.

b. Follow the instructions of part (a) with $\epsilon = 0.01$. Use $5 \le x \le 18, 1.97 \le y \le 2.03$. Copy in Figure 2 the bar graph for $j = 6, 7, 8, \dots 17$ with the lines $y = 2.01$ and $y = 1.99$.

c. Follow the instructions of part (a) with $\epsilon = 0.001$. Copy in Figure 3 the bar graph for $j = 28, 29, 30, \dots 39$ and the lines $4 = 2.001$ and $y = 1.999$ with $27 \le x \le 40, 1.997 \le y \le 2.003$.

d. Give a formula for a suitable N_ϵ as a function of $\epsilon > 0$.

Problem 2 The sequence $b_j = \dfrac{8j}{j^2 + 10}$ $(j = 1, 2, 3, \dots)$ tends to 0 as j tends to ∞ because

$$|b_j| = \dfrac{8j}{j^2 + 10} < \dfrac{8j}{j^2} = \dfrac{8}{j} \qquad (2)$$

and $8/j \to 0$. According to the ϵN-definition of this limit there is for every positive number ϵ, a positive N_ϵ such that $|b_j| < \epsilon$ for $j > N_\epsilon$.

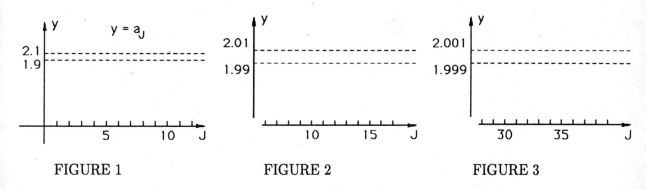

FIGURE 1   FIGURE 2   FIGURE 3

---

†See Calculator instructions 1B.3 for tips on using a calculator to solve these problems.

a. Use (2) to find an N_ϵ for $\epsilon = 1$. As a partial check of your result, use Program 6B to plot b_j for $j = 1, 2, 3, \ldots, 12$ with the line $y = \epsilon$ for $-1 \leq x \leq 13$, $-0.1 \leq y \leq 1.6$. Copy the graph in Figure 4.

b. Follow the instructions of part (a) with $\epsilon = 0.1$. Use $74 \leq x \leq 91, -0.01 \leq y \leq 0.15$; generate the plot for $j = 75, 76, 77, \ldots, 90$, and copy it in Figure 5.

c. Give a formula for a suitable N_ϵ as a function of ϵ.

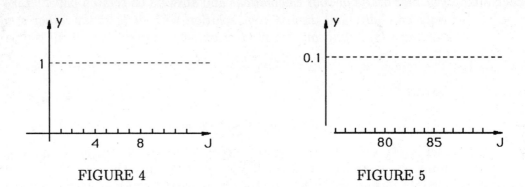

FIGURE 4   FIGURE 5

Problem 3   The sequence $c_j = \ln(j) - 1$ $(j = 1, 2, 3, \ldots)$ tends to ∞ as j tends to ∞ because $\ln(j)$ tends to ∞. According to the YN-definition of this limit there is for every positive number Y, no matter how large, a positive number N_Y such that $C_j > Y$ for $j > N_Y$.‡

a. Find an N_Y for $Y = 1$. As a partial check, use Program 6B to calculate and plot the values of c_j for $j = 1, 2, 3, \ldots, 12$ with $-1 \leq x \leq 13, -1.5 \leq y \leq 1.8$, and copy the graph in Figure 6.

b. Follow the instructions of part (a) with $Y = 10$. Use $59869 \leq x \leq 59885$, $9.9997 \leq y \leq 10.0002$. Generate the plot for $j = 59870, 59861, \ldots, 59884$ and copy it in Figure 7.

c. Give a formula for a suitable N_Y as function of $Y > 0$.

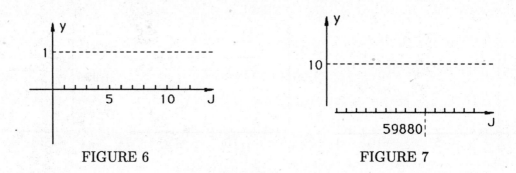

FIGURE 6   FIGURE 7

---

‡ $a_j \to -\infty$ as $j \to \infty$ if and only if for every $Y > 0$ there is an N_Y such that $a_j < -Y$ for $j > N_Y$.

Graphing calculator workbook

Series program 6D

Texas Instruments TI-81

This program calculates and plots partial sums $\sum\limits_{J=J_0}^{N} a_J$ of an infinite series $\sum\limits_{J=J_0}^{\infty} a_J$, where a_J is a function of J given as either Y_1, Y_2, Y_3, or Y_4 in the Y= menu and sutitable ranges of x and y have been chosen for the graph. The user picks Y_L, J_0, and $N = MaxJ$ when the program runs. To allow for multiple plots, the program does not clear the graphics screen. Subprogram P, which is listed with the direction field program 5B, should also be in the calculator to generate graphs of Y_1, Y_2, Y_3 or Y_4 as functions of X in the Y= menu.

The main program (205 bytes)

1.   Prgm8: SERIES
2.   : All-Off
3.   : Disp "GIVE L FOR SUMS"
4.   : Disp "OF $AJ = YL(J)$"
5.   : Input L
6.   : Disp "$J\emptyset =$"   ($\emptyset$ denotes zero.)
7.   : Input A
8.   : Disp "MAX J ="
9.   : Input N
10.   : Disp "SHOW SUMS "
11.   : Disp "AFTER N="
12.   : Input M
13.   : $\emptyset \to S$
14.   : $A \to J$
15.   : Lbl 1
16.   : If $L = 1$
17.   : $Y_1 + S \to S$
18.   : If $L = 2$
19.   : $Y_2 + S \to S$
20.   : If $L = 3$
21.   : $Y_3 + S \to S$
22.   : If $L = 4$
23.   : $Y_4 + S \to S$
24.   : If $J \leq M$
25.   : Goto 2
26.   : Disp "$N =$"
27.   : Disp J
28.   : Disp "$SN =$"
29.   : Disp S

30.   : Pause

31.   : Lbl 2

32.   : PT-On(J, S)

33.   : If $J \geq M$

34.   : Pause

35.   : $J + 1 \rightarrow J$

36.   : If $J \leq N$

37.   : Goto 1

38.   : PrgmP

Entering the main program

Press $\boxed{\text{2nd}}$ $\boxed{\text{QUIT}}$ $\boxed{\text{PRGM}}$ $\boxed{\blacktriangleright}$ and $\boxed{8}$ for program # 8.

---

1. Press $\boxed{\text{S}}$ $\boxed{\text{E}}$ $\boxed{\text{R}}$ $\boxed{\text{I}}$ $\boxed{\text{E}}$ $\boxed{\text{S}}$ $\boxed{\text{ENTER}}$.

---

2. Press $\boxed{\text{2nd}}$ $\boxed{\text{Y-VARS}}$ $\boxed{\blacktriangleleft}$ $\boxed{1}$ $\boxed{\text{ENTER}}$. All-Off unselects the functions in the Y= menu so their graphs will not be generated.

---

3. Press $\boxed{\text{PRGM}}$ $\boxed{\blacktriangleright}$ $\boxed{1}$ $\boxed{\text{2nd}}$ $\boxed{\text{A-LOCK}}$ $\boxed{\text{``}}$ $\boxed{\text{G}}$ $\boxed{\text{I}}$ $\boxed{\text{V}}$ $\boxed{\text{E}}$ $\boxed{\text{-}}$ $\boxed{\text{L}}$ $\boxed{\text{-}}$ $\boxed{\text{F}}$ $\boxed{\text{O}}$ $\boxed{\text{R}}$ $\boxed{\text{-}}$ $\boxed{\text{S}}$ $\boxed{\text{U}}$ $\boxed{\text{M}}$ $\boxed{\text{S}}$ $\boxed{\text{``}}$ $\boxed{\text{ENTER}}$.

---

4. Press $\boxed{\text{PRGM}}$ $\boxed{\blacktriangleright}$ $\boxed{1}$ $\boxed{\text{2nd}}$ $\boxed{\text{A-LOCK}}$ $\boxed{\text{``}}$ $\boxed{\text{O}}$ $\boxed{\text{F}}$ $\boxed{\text{-}}$ $\boxed{\text{A}}$ $\boxed{\text{J}}$ $\boxed{\text{2nd}}$ $\boxed{\text{TEST}}$ $\boxed{1}$ $\boxed{\text{ALPHA}}$ $\boxed{\text{Y}}$ $\boxed{\text{ALPHA}}$ $\boxed{\text{L}}$ $\boxed{(}$ $\boxed{\text{ALPHA}}$ $\boxed{\text{J}}$ $\boxed{)}$ $\boxed{\text{ALPHA}}$ $\boxed{\text{``}}$ $\boxed{\text{ENTER}}$.

---

5. Press $\boxed{\text{PRGM}}$ $\boxed{\blacktriangleright}$ $\boxed{2}$ $\boxed{\text{ALPHA}}$ $\boxed{\text{L}}$ $\boxed{\text{ENTER}}$. Here the user picks the number of the Y_L to use.

---

6. Press $\boxed{\text{PRGM}}$ $\boxed{\blacktriangleright}$ $\boxed{1}$ $\boxed{\text{ALPHA}}$ $\boxed{\text{``}}$ $\boxed{\text{ALPHA}}$ $\boxed{\text{J}}$ $\boxed{\text{0}}$ $\boxed{\text{2nd}}$ $\boxed{\text{TEST}}$ $\boxed{1}$ $\boxed{\text{ALPHA}}$ $\boxed{\text{``}}$ $\boxed{\text{ENTER}}$.

---

7. Press $\boxed{\text{PRGM}}$ $\boxed{\blacktriangleright}$ $\boxed{2}$ $\boxed{\text{ALPHA}}$ $\boxed{\text{A}}$ $\boxed{\text{ENTER}}$. Here the user assigns a value to $J0$, which is stored as A.

---

8. Press $\boxed{\text{PRGM}}$ $\boxed{\blacktriangleright}$ $\boxed{1}$ $\boxed{\text{2nd}}$ $\boxed{\text{A-LOCK}}$ $\boxed{\text{``}}$ $\boxed{\text{M}}$ $\boxed{\text{A}}$ $\boxed{\text{X}}$ $\boxed{\text{-}}$ $\boxed{\text{J}}$ $\boxed{\text{2nd}}$ $\boxed{\text{TEST}}$ $\boxed{1}$ $\boxed{\text{ALPHA}}$ $\boxed{\text{``}}$ $\boxed{\text{ENTER}}$.

---

9. Press $\boxed{\text{PRGM}}$ $\boxed{\blacktriangleright}$ $\boxed{2}$ $\boxed{\text{ALPHA}}$ $\boxed{\text{N}}$ $\boxed{\text{ENTER}}$. Set $N = MAXJ$.

Graphing calculator workbook   233

---

10. Press [PRGM] [▶] [1] [2nd] [A-LOCK] ["] [S] [H] [O] [W] [-] [S] [U] [M] [S] ["] [ENTER].

---

11. Press [PRGM] [▶] [1] [2nd] [A-LOCK] ["] [A] [F] [T] [E] [R] [-] [N] [2nd] [TEST] [1] [ALPHA] ["] [ENTER].

---

12. Press [PRGM] [▶] [2] [ALPHA] [M] [ENTER].

---

13. Press [∅] [STO▶] [S] [ENTER]. The partial sums are stored as S, which is set equal to 0 initially to erase its previous value.

---

14. Press [ALPHA] [A] [STO▶] [J] [ENTER] to set J equal to $A = J0$.

---

15. Press [PRGM] [1] [1] [ENTER]. The operation of the program will return to this label when it reaches the Goto 1 that will be put on line 37.

---

16. Press [PRGM] [3] [ALPHA] [L] [2nd] [TEST] [1] [1] [ENTER].
    If $L = 1$, the next line is executed and Y_1 is used.

---

17. Press [2nd] [Y-VARS] [1] [+] [ALPHA] [S] [STO▶] [S] [ENTER]
    to add $Y_1(J)$ to S.

---

18. Press [PRGM] [3] [ALPHA] [L] [2nd] [TEST] [1] [2] [ENTER].
    If $L = 2$, the next line is executed and Y_2 is used.

---

19. Press [2nd] [Y-VARS] [2] [+] [ALPHA] [S] [STO▶] [S] [ENTER]
    to add $Y_2(J)$ to S.

---

20. Press [PRGM] [3] [ALPHA] [L] [2nd] [TEST] [1] [3] [ENTER].
    If $L = 3$, the next line is executed and Y_3 is used.

---

21. Press [2nd] [Y-VARS] [3] [+] [ALPHA] [S] [STO▶] [S] [ENTER]
    to add $Y_3(J)$ to S.

---

22. Press [PRGM] [3] [ALPHA] [L] [2nd] [TEST] [1] [4] [ENTER].
    If $L = 4$, the next line is executed and Y_4 is used.

---

23. Press [2nd] [Y-VARS] [4] [+] [ALPHA] [S] [STO▶] [S] [ENTER] to add $Y_4(J)$ to S.

24. Press [PRGM] [3] [ALPHA] [J] [2nd] [TEST] [6] [ALPHA] [M] [ENTER]. If $J \leq M$, the next line is executed and lines which display the values of the sums are skipped.

25. Press [PRGM] [2] [2] [ENTER].

26. Press [PRGM] [▶] [1] [ALPHA] ["] [ALPHA] [N] [2nd] [TEST] [1] [ALPHA] ["] [ENTER]. Display $N =$.

27. Press [PRGM] [▶] [1] [ALPHA] [J] [ENTER]. Display the last value of J.

28. Press [PRGM] [▶] [1] [2nd] [A-LOCK] ["] [S] [N] [2nd] [TEST] [1] [ALPHA] ["] [ENTER] to have the program display $SN =$.

29. Press [PRGM] [▶] [1] [ALPHA] [S] [ENTER]. Display SN.

30. Press [PRGM] [6] [ENTER]. The program pauses to show the values.

31. Press [PRGM] [1] [2] [ENTER].

32. Press [2nd] [DRAW] [3] [ALPHA] [J] [ALPHA] [,] [ALPHA] [,] [ALPHA] [S] [ )] [ENTER]. Put a dot at (J, S) to represent the sum to $J = N$.

33. Press [PRGM] [3] [ALPHA] [J] [2nd] [TEST] [4] [ALPHA] [M] [ENTER].

34. Press [PRGM] [6] [ENTER]. The program pauses to show the graph if $J \geq M$.

35. Press [ALPHA] [J] [+] [1] [STO▶] [J] [ENTER] to increase the value of J by 1.

Graphing calculator workbook

36. Press [PRGM] [3] [ALPHA] [J] [2nd] [TEST] [6] [ALPHA] [N] [ENTER]. If J is $\leq N =$ MaxJ, the next line is executed and the loop is repeated. Otherwise the next line is skipped and the summation ends.

37. Press [PRGM] [2] [1] [ENTER].

38. Press [PRGM] [◄] [ALPHA] [P] [ENTER]. This command begins the execution of subprogram P.

Using the program

Example 1a Use the series program to display and plot the first 15 partial sums of

$$\sum_{j=1}^{\infty} \frac{(-1)^{j+1}}{j^{1.2}}.$$

b. Generate a graph of the first 100 partial sums and display the 98th through the 100th.

Solution Press [Y =] and enter the formula for the terms of the series as $Y_1 = (-1) \wedge (J+1)/J \wedge 1.2$.

a. Set x-min $= -1$, x-max $= 16$, y-min $= -0.3$ and y-max $= 1.3$ for the graph. (This y-range was chosen after some experimentation.) Press [2nd] [QUIT] [2nd] [DRAW] [1] [ENTER] to clear the graphics screen if do not change the ranges. To run the program, press [PRGM], its number, and [ENTER]. Press [1] [ENTER] after the prompt "GIVE L FOR SUMS OF $AJ = YL(j)$" to use Y_1. (You would enter 2 to use Y_2, etc.) Press [1] [ENTER] after the prompt "J0 = ?" to have the sums start at $j = 1$. Press [1] [5] [ENTER] after the prompt "MAXJ = ?" and [0] [ENTER] after "SHOW SUMS AFTER $N =$" to calculate, display, and plot the sums starting with $N = 1$ and ending with $N = 15$. Press [ENTER] at each step in the program. At the end press [0] [ENTER], indicating that you do not want any graphs of other functions, or [ON] [2] (QUIT) to terminate the program directly. Press [GRAPH] to see the graph. You should get the values below and the graph in Figure 1—with smaller dots.

$$s_1 = 1, \quad s_2 \doteq 0.5647247184, \quad s_3 \doteq 0.8323052389$$
$$s_4 \doteq 0.6428406681, \quad s_5 \doteq 0.7877966009, \quad s_6 \doteq 0.6713254144$$
$$s_7 \doteq 0.7681269735, \quad s_8 \doteq 0.6856577290, \quad s_9 \doteq 0.7572570640$$
$$s_{10} \doteq 0.6941613296, \quad s_{11} \doteq 0.7504380496, \quad s_{12} \doteq 0.6997410212$$
$$s_{13} \doteq 0.7457950870, \quad s_{14} \doteq 0.7036597611, \quad s_{15} \doteq 0.7424471450$$

b. Press $\boxed{\text{RANGE}}$ and set x-min $= -5$ and x-max $= 105$. This clears the graphics screen. Then run the program with $L = 1$, $J0 = 1$, and $\text{Max}J = 100$. Give 97 after the last prompt so the partial sums for $N = 98, 99$ and 100 will be displayed. The result is the graph in Figure 2 and the values

$$s_{98} \doteq 0.7218002999, \quad s_{99} \doteq 0.7258296757, \quad s_{100} \doteq 0.7218486040$$

The partial sums seem to have a limit between 0.721 and 0.726. This limit is the sum of the infinite series.

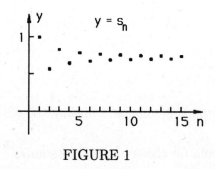
FIGURE 1

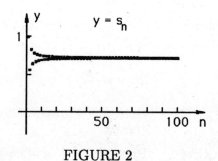
FIGURE 2

Graphing calculator workbook

Worksheet 6E.1†

The Comparison test for series with nonnegative terms

Name _____ Date _____

Others in your group _____

Instructor, Teaching Assistant, and/or Recitation section _____

Directions Put first drafts of your calculations and answers on scratch paper. Take your time, work carefully, and discuss your solution with at least one other student before putting a final draft on this sheet or on other paper. Turn in all your work.

Problem 1a Use series program 6D to plot the partial sums $s_n = \sum_{j=0}^{n} \frac{1}{j+1}(0.6)^j$ of the infinite series $\sum_{j=0}^{\infty} \frac{1}{j+1}(0.6)^j$ for $n = 0, 1, \ldots, 12$. Use $-1 \leq x \leq 13$, $-0.5 \leq y \leq 2.75$ and copy the graph in Figure 1.

b. Because the terms $\frac{1}{j+1}(0.6)^j$ of $\sum_{j=0}^{\infty} \frac{1}{j+1}(0.6)^j$ are all ≥ 0, this infinite series either converges or diverges to ∞. Prove that it converges or diverges by comparing it with the geometric series $\sum_{j=0}^{\infty} (0.6)^j$. To illustrate your reasoning, use the series program to plot together the partial sums of both series for $n = 0, 1, 2, \ldots, 12$ with the ranges of part (a). Copy the graphs in Figure 2.

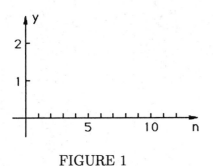

FIGURE 1

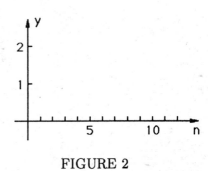

FIGURE 2

---

†This worksheet uses the Comparison test with geometric and harmonic series. See Calculator instructions 1B.3 for tips on working it with a calculator.

Problem 2 Determine whether the infinite series $\sum_{j=1}^{\infty} \frac{1}{j^{3/4}}$ converges or diverges to ∞ by comparing it with the harmonic series $\sum_{j=1}^{\infty} \frac{1}{j}$. To illustrate your reasoning, use the series program to generate graphs of partial sums of both series for $n = 1, 2, 3, \ldots, 12$. Use $-1 \le x \le 13$, $-0.5 \le y \le 5$ and copy the graphs in Figure 3.

Problem 3 Determine whether the infinite series $\sum_{j=0}^{\infty} \frac{4 + \sin j}{2^j}$ converges or diverges by comparing it with the geometric series $\sum_{j=0}^{\infty} \frac{5}{2^j}$. To illustrate your reasoning, generate graphs of partial sums of both series for $n = 0, 1, 2, \ldots, 12$. Use $-1 \le x \le 13$, $-1 \le y \le 11$ and copy the graphs in Figure 4.

Problem 4 Use the series program to plot the partial sums of $\sum_{j=1}^{\infty} \frac{\ln j}{5 \ln j - 1}$ for $n = 1, 2, 3, \ldots, 20$. Use $-1 \le x \le 21$, $-0.5 \le y \le 5$ and copy the graph in Figure 5. Does this series converge or diverge? Justify your conclusion

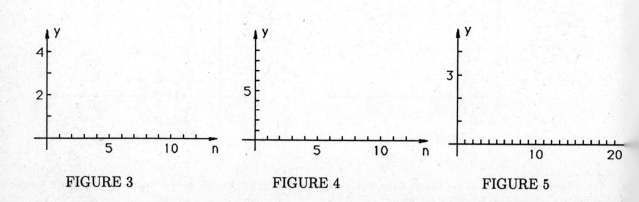

FIGURE 3   FIGURE 4   FIGURE 5

Worksheet 6E.2[†]

Alternating series

Name _____ Date _____

Others in your group _____

Instructor, Teaching Assistant, and/or Recitation section _____

Directions Put first drafts of your calculations and answers on scratch paper. Take your time, work carefully, and discuss your solution with at least one other student before putting a final draft on this sheet or on other paper. Turn in all your work.

Problem 1a Use series program 6D to list and plot the partial sums of $s_n = \sum_{j=0}^{n} (-0.9)^j$ for $n = 0, 1, \ldots, 11$ of $\sum_{j=0}^{\infty} (-0.9)^j$. Use $-1 \leq x \leq 12$, $-0.2 \leq y \leq 1.3$. Complete the table below of values of the partial sums below and copy the graph in Figure 1. Notice that the partial sums for even n decrease and those for odd n increase.

  b. Generate a graph of the partial sums for $n = 50, 51, 52, \ldots, 62$ without listing the values and copy it in Figure 2. Use $49 \leq x \leq 63$, $0.516 \leq y \leq 0.533$, and y-scale $= 0.002$. (It may take a calculator several seconds to calculate the first 50 sums.)

  c. Use the Alternating series test to show that the infinite series converges or prove that it diverges.

s_0	1	s_1	0.1
s_2	0.91	s_3	1.81
s_4		s_5	
s_6		s_7	
s_8		s_9	
s_{10}		s_{11}	

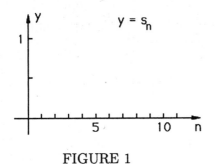

FIGURE 1

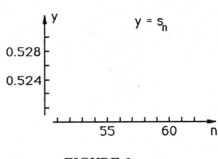

FIGURE 2

[†]This worksheet uses the Alternating series test. See Calculator instructions 1B.3 for tips on solving it with a calculator.

Problem 2 Generate graphs of the partial sums $s_n = \sum_{j=1}^{n} \frac{(-1)^{j+1}}{j}$ of $\sum_{j=1}^{\infty} \frac{(-1)^{j+1}}{j}$ first for $n = 1, 2, 3, \ldots, 20$, using $-1 \leq x \leq 21, -0.25 \leq y \leq 1.25$, y-scale $= 0.5$, and then for $n = 101, 102, 103, \ldots, 120$, using $101 \leq x \leq 121, 0.65 \leq y \leq 0.72$, y-scale $= 0.01$. Copy the graphs in Figures 3 and 4. Then use the Alternating series test to show that the infinite series converges or prove that it diverges.

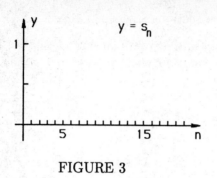
FIGURE 3

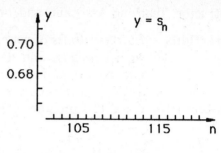
FIGURE 4

Problem 3 Generate graphs of the partial sums $s_n = \sum_{j=0}^{n} (-1)^j \frac{j+5}{j+1}$ of $\sum_{j=0}^{\infty} (-1)^j \frac{j+5}{j+1}$ first for $n = 0, 1, 2, \ldots, 20$, using $-1 \leq x \leq 21, -1 \leq y \leq 6$, and then for $n = 101, 102, 103, \ldots, 120$, using $101 \leq x \leq 121, -1 \leq y \leq 6$. Copy the graphs in Figures 5 and 6. Use the Alternating series test to show that the infinite series converges or prove that it diverges.

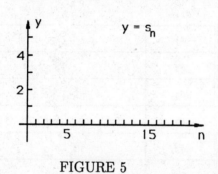
FIGURE 5

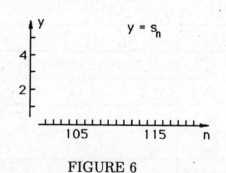
FIGURE 6

Problem 4 Generate a graph of the partial sums $s_n = \sum_{j=1}^{n} \frac{1-(-1)^j}{\sqrt{j}}$ of $\sum_{j=1}^{\infty} \frac{1-(-1)^j}{\sqrt{j}}$ for $n = 1, 2, 3, \ldots, 12$ and copy it in Figure 7. Show that the infinite series diverges. Why does the Alternating series test not apply?

FIGURE 7
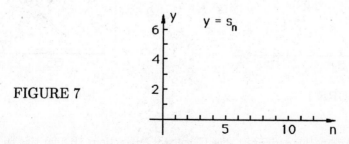

Graphing calculator workbook

Worksheet 6E.3[†]

Using convergence and divergence tests

Name _____ Date _____

Others in your group _____

Instructor, Teaching Assistant, and/or Recitation section _____

Directions Put first drafts of your calculations and answers on scratch paper. Take your time, work carefully, and discuss your solution with at least one other student before putting a final draft on this sheet or on other paper. Turn in all your work.

In each of the following problems, use the series program 6D to plot the partial sums $s_n = \sum_{j=j_0}^{n} a_j$ of the infinite series for $n = j_0, \ldots, 20$ and to display the values of the last four partial sums, s_{17}, s_{18}, s_{19} and s_{20}. Use $-1 \leq x \leq 22$ and the y-range given with problem. Copy the graph and list the approximate decimal values of the partial sums.

Then determine whether the series converges, diverges to ∞, diverges to $-\infty$, or diverges by oscillation and justify your conclusion with a convergence or divergence test.

Problem 1 $\displaystyle\sum_{j=0}^{\infty} \frac{\cos(j^2)}{j^3 + 1}$ $(-0.25 \leq y \leq 1.75)$

Problem 2 $\displaystyle\sum_{j=2}^{\infty} \frac{(-1)^{j+1}}{\ln(\ln j)}$ $(j_0 = 2, -2 \leq y \leq 18)$

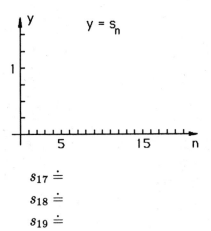

FIGURE 1 FIGURE 2

[†]This worksheet uses the Comparison, Limit comparison, Alternating series, and Ratio tests. See Calculator instructions 1B.3 for tips on solving it with a calculator.

Problem 3 $\quad \sum_{j=2}^{\infty} \dfrac{1}{\ln(\ln j)} \quad (j_0 = 2, -10 \leq y \leq 35))$

Problem 4 $\quad \sum_{j=1}^{\infty} \left(\dfrac{8}{j}\right)^j \quad (j_0 = 1, -10 \leq y \leq 100))$

Problem 5 $\quad \sum_{j=0}^{\infty} (-1)^j \arctan(j) \quad (-1.5 \leq y \leq 1.5)$

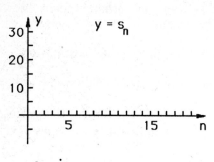
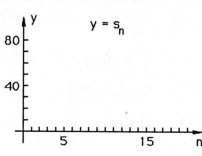
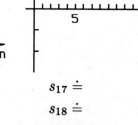

$s_{17} \doteq$
$s_{18} \doteq$
$s_{19} \doteq$
$s_{20} \doteq$

$s_{17} \doteq$
$s_{18} \doteq$
$s_{19} \doteq$
$s_{20} \doteq$

$s_{17} \doteq$
$s_{18} \doteq$
$s_{19} \doteq$
$s_{20} \doteq$

FIGURE 3   FIGURE 4   FIGURE 5

Problem 6 $\quad \sum_{j=0}^{\infty} \dfrac{j^4}{j!} \quad (-10 \leq y \leq 60)$

Problem 7 $\quad \sum_{j=1}^{\infty} \dfrac{j!}{j^{12}} \quad (j_0 = 1, -0.5 \leq y \leq 3))$

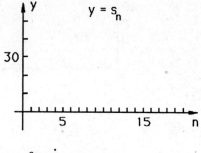
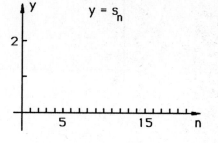

$s_{17} \doteq$
$s_{18} \doteq$
$s_{19} \doteq$
$s_{20} \doteq$

$s_{17} \doteq$
$s_{18} \doteq$
$s_{19} \doteq$
$s_{20} \doteq$

FIGURE 6   FIGURE 7

Graphing calculator workbook

Worksheet 6F.1†

Matching functions to Taylor polynomial approximations

Name _____ Date _____

Others in your group _____

Instructor, Teaching Assistant, and/or Recitation section _____

Directions Put first drafts of your calculations and answers on scratch paper. Take your time, work carefully, and discuss your solution with at least one other student before putting a final draft on this sheet or on other paper. Turn in all your work.

Problem 1a Match the functions

$$f_1 = 2xe^{-x}, \qquad f_2 = \frac{1}{1-x}, \qquad f_3 = \ln(1-x)$$

to the graphs in Figures 1 through 3 without using your calculator or computer and put the formulas under the corresponding graphs. Then check your conclusions by generating the curves on your calculator or computer with $-2 \leq x \leq 3, -3 \leq y \leq 3$.

b. Generate the graphs of each of the following polynomials with the ranges of x and y from part (a) and determine which of the curves in Figures 1 through 3 each one best approximates near $x = 0$. Regenerate it with the graph it approximates, copy it in the corresponding figure, and put its formula under the graphs.

$$Q_1 = 1 + x + x^2 + x^3, \qquad Q_2 = 2x - 2x^2 + x^3, \qquad Q_3 = -x - \tfrac{1}{2}x^2 - \tfrac{1}{3}x^3$$

c. Show that each of the polynomials $Q_1, Q_2,$ and Q_3 has the same value and the same first three derivatives at $x = 0$ as the corresponding function from part (a).

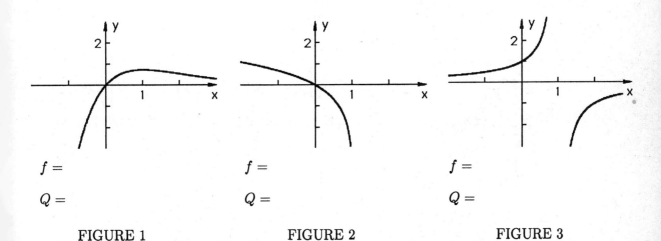

FIGURE 1   FIGURE 2   FIGURE 3

---

†This worksheet is an introduction to Taylor polynomials.

Problem 2a Match the functions

$$f_4 = \cos x, \qquad f_5 = x \cos x, \qquad f_6 = \sin x$$

to the graphs in Figures 4 through 6 without using your calculator or computer and put the formulas under the corresponding graphs. Check your conclusions by generating the curves on your calculator or computer with $-6 \leq x \leq 6, -4 \leq y \leq 4$.

b. Generate the graphs of each of the following polynomials with the ranges from part (a) and determine which of the curves in Figures 4 through 6 each one best approximates near $x = 0$. Regenerate it with the graph it approximates, copy it in the corresponding figure, and put its formula under the graphs.

$$Q_4 = x - \frac{1}{2}x^3 + \frac{1}{4!}x^5, \qquad Q_5 = x - \frac{1}{3!}x^3 + \frac{1}{5!}x^5, \qquad Q_6 = 1 - \frac{1}{2}x^2 + \frac{1}{4!}x^4$$

c. Show that each of the polynomials Q_4, Q_5, and Q_6 has the same value and the same first five derivatives at $x = 0$ as the corresponding function from part (a).

Problem 3 Polynomials Q_1 through Q_6 are called Taylor polynomial approximations of the corresponding functions f_1 through f_6. What general principle about Taylor polynomials do the results of Problems 1c and 2c suggest?

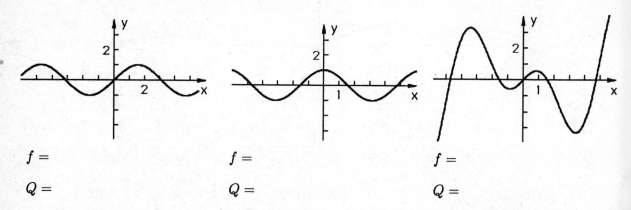

FIGURE 4   FIGURE 5   FIGURE 6

Graphing calculator workbook

Worksheet 6F.2[†]

Adding more terms; Looking near and far away

Name _____ Date _____

Others in your group _____

Instructor, Teaching Assistant, and/or Recitation section _____

Directions Put first drafts of your calculations and answers on scratch paper. Take your time, work carefully, and discuss your solution with at least one other student before putting a final draft on this sheet or on other paper. Turn in all your work.

In these problems $P_n(x)$ denotes an nth degree Taylor polynomial approximation centered at $x = 0$ of a function $f(x)$ with n derivatives. P_n is given by

$$P_n(x) = f(0) + f'(0)x + \frac{1}{2!}f''(0)x^2 + \frac{1}{3!}f^{(3)}(0)x^3 + \cdots \frac{1}{n!}f^{(n)}(0)x^n. \qquad (1)$$

Problem 1 Figure 1 shows the graphs of $\cos x$ and its Taylor polynomial $P_0(x) = 1$. The first row of Table shows the values of $\cos(1)$, $P_0(1)$, and $\cos(1) - P_0(1)$. Find $P_2(x)$; generate its graph with the graph of $\cos x$ for $-6 \le x \le 6, -4 \le y \le 4$; copy it in Figure 2; and put the values of $P_2(1)$ and $\cos(1) - P_2(1)$ on the second row of the table, with three digit accuracy in column 3. Then repeat the process with $P_4, P_6, P_8,$ and P_{10}, using Figures 3 through 6 and the other rows of the table. How does increasing n affect the approximations?

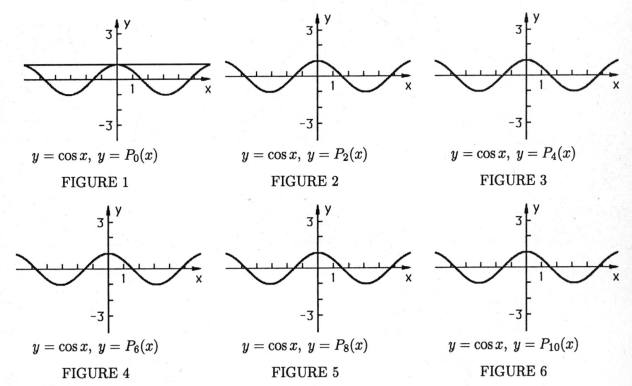

[†]These problems use the formula for Taylor polynomials that is given in the worksheet. See Calculator instructions 1B.3 for tips on working it with a calculator.

n	$\cos(1)$	$P_n(1)$	$\cos(1) - P_n(1)$
1	0.5403023050	1	-0.460
2	0.5403023050		
4	0.5403023050		
6	0.5403023050		2.45×10^{-5}
8	0.5403023050		
10	0.5403023050	0.5403023038	2.08×10^{-9}

TABLE 1

Problem 2 Find the Taylor polynomial approximation $P_3(x)$ of $5 + e^x$, centered at $x = 0$. Generate the graphs of $e^x + 5$ and $P_3(x)$ with $-5 \leq x \leq 5, -10 \leq y \leq 25$, and y-scale $= 5$ and copy them in Figure 7. Then complete Table 2 of values of the two functions, with three digit accuracy in the last column. What do the graph and table suggest about the approximation for values of x that are close or far from 0?

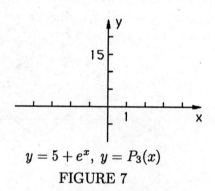

$y = 5 + e^x$, $y = P_3(x)$

FIGURE 7

x	$5 + e^x$	$P_3(x)$	$5 + e^x - P_3(x)$
1	7.718281828	7.666666667	5.16×10^{-2}
0.1			
0.01			
2	12.3890561	11.33333333	1.06
3			
15			

TABLE 2

Graphing calculator workbook

Worksheet 6F.3†

Error estimates vs. actual errors

Name _____ Date _____

Others in your group _____

Instructor, Teaching Assistant, and/or Recitation section _____

Directions Put first drafts of your calculations and answers on scratch paper. Take your time, work carefully, and discuss your solution with at least one other student before putting a final draft on this sheet or on other paper. Turn in all your work.

If $f(x)$ has $n+1$ continuous derivatives in an open interval I containing 0, then for x in I, the error in approximating $f(x)$ by its nth degree Taylor polynomial $P_n(x)$ centered at 0 satisfies

$$|f(x) - P_n(x)| \leq \frac{1}{(n+1)!} M_{n+1} |x|^{n+1} \tag{1}$$

where M_{n+1} is the maximum of $|f^{n+1}(t)|$ for $0 \leq t \leq x$ if $x \geq 0$ and for $x \leq t \leq 0$ if $x < 0$.

Problem 1a Find $P_6(x)$ for $\sin x$ and generate its graph with the graph of $\sin x$ on your calculator or computer for $-6 \leq x \leq 6, -4 \leq y \leq 4$. Copy the graphs in Figure 1. Notice that in this case $P_5 = P_6$.

b. Show that the error estimate (1) in this case gives

$$|\sin x - P_6(x)| \leq \frac{1}{7!} |x|^7. \tag{2}$$

c. Calculate $\sin x - P_6(x)$ and $\frac{1}{7!} |x|^7$ at $x = 0$ and $x = 1$. You should get the numbers in the first two rows of Table 1. Complete the table with three digit accuracy by finding the values of the error and its estimate at $x = 0.5, 0.1, -3, 5,$ and -10. What does the table indicate about the error estimates as compared with actual errors?

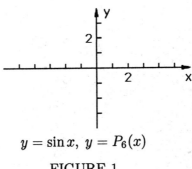

$y = \sin x, \ y = P_6(x)$

FIGURE 1

---

†These problems use an error estimate that is given in the worksheet. See Calculator instructions 1B.3 for tips on solving it with a calculator.

x	$\sin x - P_6(x)$	$\frac{1}{7!}|x|^7$
0	0	0
1	-1.96×10^{-4}	1.98×10^{-4}
0.5		
0.1		
-3		
5		
-10		

TABLE 1

Problem 2a Find P_2 for $\ln(1+x)$. Generate the graphs of both functions for $-1.5 \leq x \leq 3$, $-2 \leq y \leq 3$ and copy them in Figure 2.

    b. Show that for $x \geq 0$ the error estimate (1) gives, for $x \geq 0$,

$$|\ln(1+x) - P_2(x)| \leq \tfrac{1}{3}|x|^3. \qquad (2)$$

    c. Use (2) to find a number b such that $|\ln(1+x) - P_2(x)| \leq 1$ for $0 \leq x \leq b$.

    d. Generate $y = |\ln(1+x) - P_2(x)|$ and $y = 1$ with the same ranges as in part (a) and copy them in Figure 3. Use the graphs to determine the actual approximate range of positive x such that $|\ln(1+x) - P_2(x)| \leq 1$.

    e. Explain the logical relationship between the answers to parts (c) and (d).

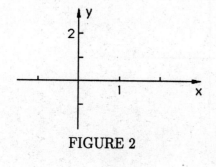

FIGURE 2

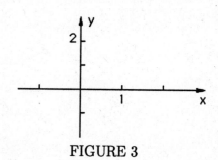

FIGURE 3

Graphing calculator workbook

Worksheet 6F.4[†]

Operations with power series

Name _____ Date _____

Others in your group _____

Instructor, Teaching Assistant, and/or Recitation section _____

Directions Put first drafts of your calculations and answers on scratch paper. Take your time, work carefully, and discuss your solution with at least one other student before putting a final draft on this sheet or on other paper. Turn in all your work.

The problems on this worksheet use the following power series:

$$\sin x = x - \frac{1}{3!}x^3 + \frac{1}{5!}x^5 - \frac{1}{7!}x^7 + \cdots \qquad (1)$$

$$e^x = 1 + x + \frac{1}{2!}x^2 + \frac{1}{3!}x^3 + \frac{1}{4!}x^4 + \cdots \qquad (2)$$

$$\frac{1}{1-x} = 1 + x + x^2 + x^3 + x^4 + x^5 + \cdots \qquad (3)$$

Problem 1 Use (1) to find the Taylor polynomial $P_{10}(x)$ for $\sin(x^2)$ centered at 0. As a partial check on your result, generate the graphs of $\sin(x^2)$ and $P_{10}(x)$ for $-3 \le x \le 3$, $-2 \le y \le 4$ and copy them in Figure 1. Label the curves to distinguish them. As a further check, compare the values of the two functions at some very small, nonzero values of x.

Problem 2 Use (1) to find the Taylor polynomial $P_4(x)$ for $\dfrac{\sin x}{x}$ centered at 0. As a partial check on your result, generate the graphs of $\dfrac{\sin x}{x}$ and $P_4(x)$ for $-8 \le x \le 8$, $-1 \le y \le 2.5$. Copy and label them in Figure 2. As a further check, compare the values of the two functions at some small values of x.

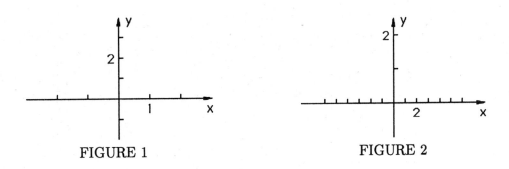

FIGURE 1 FIGURE 2

[†]This worksheet involves finding Taylor series of new functions by operating on known Taylor polynomials of other functions, and shows how such resuts can be partially checked with graphs and function evaluations. See Calculator instructions 1B.3 for tips on working it with a calculator.

Problem 3 Use (1) to find the Taylor polynomial $P_6(x)$ for $(\sin x)^2$ centered at 0. (Be sure P_6 has no terms of degree > 6.) As a partial check on your result, generate the graphs of $(\sin x)^2$ and $P_6(x)$ for $-4 \leq x \leq 4$, $-1 \leq y \leq 2.5$. Copy them in Figure 3.

Problem 4 Use (2) to find the Taylor polynomial $P_3(x)$ for $\frac{1}{2}(e^x - e^{-x})$ centered at 0. As a partial check on your result, generate the graphs of $\frac{1}{2}(e^x - e^{-x})$ and $P_4(x)$ for $-4 \leq x \leq 4$, $-30 \leq y \leq 30$ with y-scale $= 10$. Copy and label them in Figure 4.

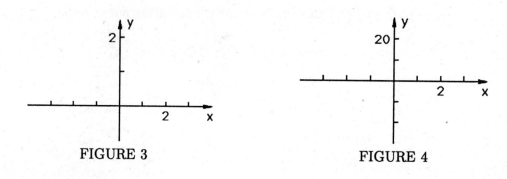

FIGURE 3 FIGURE 4

Problem 5 Use (3) to find the Taylor polynomial $P_3(x)$ for $\dfrac{d}{dx}\left[\dfrac{1}{1-x}\right] = \dfrac{1}{(1-x)^2}$ centered at 0. As a partial check on your result, generate the graphs of $\dfrac{1}{(1-x)^2}$ and $P_3(x)$ for $-2 \leq x \leq 2$, $-2 \leq y \leq 8$. Copy and label them in Figure 5.

Problem 6 Use (3) with x replaced by $-t^2$ to find the Taylor polynomial $P_7(x)$ for $\arctan x = \displaystyle\int_0^x \dfrac{1}{1+t^2}\,dt$ centered at 0. As a partial check on your result, generate the graphs of $\arctan x$ and $P_7(x)$ for $-3 \leq x \leq 3$, $-2.5 \leq y \leq 2.5$. Copy and label them in Figure 6.

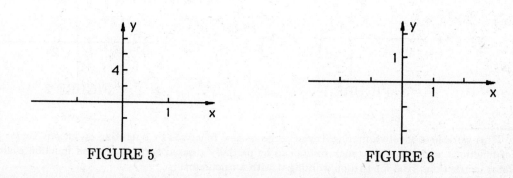

FIGURE 5 FIGURE 6

Graphing calculator workbook   p. 251

INDEX

Alternating series, 239–240
Calculator instructions
    basic, 9–12
    generating graphs,13–14
    pixels and angles, 23–24
    substitute pages of, v, vii
    tips for solving worksheets, 15–18
Calculators
    Casio,v, vii
    selection, v, x
    Sharp, v, vii
    Texas Instruments, v, vii
Comparison test for infinite series, 237–238
Compound interest, 183–184
Computers, v, x
Constructions, v
    solids of revolution, 161–166
    volumes by slicing, 157–160
Continuity, 53–54
Convergence and divergence of sequences, 229-230
Cooperative learning, v, vi
Course outlines, v
    for regular courses, xi
    for short courses, xvii
Course planning, v–vii
Derivative tests
    first, 97–98
    second, 99–100
Derivatives
    a closer look, 87–88
    approximate, graphs of, 89–90
    approximate, from graphs, 69–72, 81–82
    generating graphs of, 89–90
    of composite functions, 83–86
    numerical experiments, 77–80, 173–176
Differential equations
    Direction field program, 199–206
    Euler's method, 207–208
    of families of curves, 197–198
    of growth and decay, 193–194
    of velocity with resistance, 195–196
    Runge-Kutta program, 209–216
Equations and inequalities, 39–46
Euler's method, 207–208

Exponential functions
    (see Logarithms and exponential functions)
Extreme value theorem, 53–54
First derivative test, 97–98
Functions
    domains of, 45–46
    graphs and values of, 25–30
Fundamental theorem
    derivatives of integrals, 147–148
    integrals of derivatives, 149–150
Geometric series
    and probability, 221–222
    Going part way, 217–218
    Loan payments and St. Ives, 219–220
Graph sketching
    with calculus, 97–100
    with sines and cosines, 101–102
    with vertical tangent lines, 87–88, 103–104
    without calculus, 91–96
Group work, vi
Instructor's manual (*Graphs and Answers*), vii
Integrals
    and average values, 133–134
    and areas, 129–134
    and lengths of curves, 133-134
    and velocity, 129–130
    and volumes, 131–132, 157–166
    initial investigations, 129–134
    numerical experiments, 141–146
    of $1/x$, 177–178
    of linear combinations, 145–146
    of powers, 141–142
    of rational functions, 191–192
    of sines and cosines, 143–144
    Riemann sum program, 135–140
    Simpson's rule program, 151–156
Intermediate value theorem, 53–54
Inverse trigonometric functions
    applications, 185–187
    functions involving, 189–190
Limits
    infinite, 51–52
    involving trigonometric functions, 59–60
    one-sided, finite, 49–50
    two-sided finite, 47–48

Logarithms and exponential functions
  applications, 179–182
  algebra of, 65–66
  derivatives of, 173–176
  equations involving, 63–64
  graphs involving, 61–62
  integrals of 1/x, 177–178
Maximum/minimum problems
  a graphical approach, 31–38
  initial investigations, 1–8
  with inverse trig. functions,185–188
Newton's method
  applications, 121–128
  Newton's method program, 115–120
Partial fractions, 191–192
Power series, 243–250
Probability, 221–222
Programs
  Direction field, 199–205
  Masterpiece around the corner, 185
  Newton's method, 115–120
  Riemann sum, 135–140
  Runge-Kutta , 209–216
  Secant line, 73–76
  Sequence, 223–226
  Series, 231–236
  Simpson's rule, 151–156
  What's the angle?, 187
Projects using integrals, v
  Billy the kid, 169–170
  designing a boat, 171
  designing a shish-kabob, 167
Rates of change
  of a height, 111-112
  of a length, 113–114
  of linear functions, 105–106
  of a distance, 107–108
  of volumes, 109–110
Riemann sum program, 135–140
Riemann sums, using, 141–146, 177–178
Runge-Kutta program, 209–216
Secant line program, 73–76
Second derivative test, 99-100
Sequence program, 223–226
Sequences
  convergence and divergence, 229–230
  initial investigations, 227–28
  Sequence program 223–226

Series
  alternating series, 239–240
  Comparison test, 237–238
  Series program, 231–236
  operations with power series, 249–250
  using convergence tests, 241–242
  Series program, 231–236
Simpson's rule program, 151–156
Student study guide (to Shenk's *Calculus*), vii
Substitute pages, v, vii
Tangent lines
  numerical experiments, 77–80, 173–176
  Secant line program, 73–76
  vertical, 103–104
Taylor polynomials
  errors with, 245–248
  matching to functions, 243–244
Teaching philosophy, v–vii
Textbooks, v, vii
Trigonometric functions
  definitions of, 55–56
  derivatives of, 79–80
  equations involving, 57–58
  limits involving, 59–60
Velocity
  nonconstant, graphical approach, 69–72
  piecewise constant, 67–68
Vertical tangent lines, 87–88, 103–104
Worksheets
  exploratory, v
  use by students, ix–x
Workshop sessions, vi

Notes

Notes

Notes

Notes

Notes

Notes

Notes